Biofuels

Production, Application and Development

Biofuels

Production, Application and Development

A.H. Scragg

www.cabi.org

CABI is a trading name of CAB International

CABI Head Office
Nosworthy Way
Wallingford
Oxfordshire OX10 8DE
UK

CABI North American Office
875 Massachusetts Avenue
7th Floor
Cambridge, MA 02139
USA

Tel: + 44 (0)1491 832111
Fax: + 44 (0)1491 833508
E-mail: cabi@cabi.org
Website: www.cabi.org

Tel: + 1 617 395 4056
Fax: + 1 617 354 6875
E-mail: cabi-nao@cabi.org

A catalogue record for this book is available from the British Library, London, UK.

Library of Congress Cataloging-in-Publication Data

Scragg, A.H. (Alan H.), 1943-

Biofuels, production, application and development/A.H. Scragg.
   p. cm.
  Includes bibliographical references and index.
  ISBN 978–1–84593–592–4 (alk. paper)
1. Biomass energy. 2. Renewable energy sources. 3. Biomass energy--Environmental aspects.
4. Renewable energy sources--Environmental aspects. I. Title.

HD9502.5.B542S35 2009
333.95'39--dc22

2009012259

ISBN-13: 978 1 84593 592 4

Typeset by SPi, Pondicherry, India.
Printed and bound in the UK by Cambridge University Press, Cambridge.

The paper used for the text pages in this book is FSC certified. The FSC (Forest Stewardship Council) is an international network to promote responsible management of the world's forests.

Contents

Preface

Biofuels are energy sources derived from biological materials, which distinguishes them from other non-fossil fuel energy sources such as wind and wave energy. Biofuels can be solid, liquid or gaseous, and all three forms of energy are sustainable and renewable because they are produced from plants and animals, and therefore can be replaced in a short time span. In contrast, fossil fuels have taken from 10 to 100 million years to produce and what we are burning is ancient solar energy. In addition, the energy derived from plant material should be intrinsically carbon-neutral, as the carbon accumulated in the plants by the fixation of carbon dioxide in photosynthesis is released when the material is burnt.

At present it is clear that the supply of fossil fuels is finite and a time can be envisaged when the supplies of fossil fuels become scarce or even run out. Also the burning of fossil fuels releases additional carbon dioxide into the atmosphere over and above that released in the normal carbon cycle. The accumulation of carbon dioxide in the atmosphere appears to be the major cause of global warming. The consensus suggests that the long-term effects of global warming will be severe, with drastic changes in climate and sea levels. At the same time, modern society requires increasing amounts of energy, most of which is obtained from fossil fuels. Thus, mankind is almost totally reliant on fossil fuels to provide electricity, heating/cooling and transport fuels. This reliance can be seen in the effects on countries when oil supplies are interrupted by wars, embargos and strikes. In 2008, the world suffered from rapid rises in oil prices which affected the price of many commodities.

Alternative energy supplies are needed, therefore, to provide both power and fuel for transport. The possible energy sources available are very diverse and include hydroelectric, nuclear, wind, biological materials and many others. Whatever energy source is used, it should be sustainable and as carbon-neutral as possible. Biofuels encompass the contribution that biological materials may make to energy supply and in particular liquid fuels for transport. Solid biofuels, principally biomass, have been used for thousands of years to provide heat and for cooking, and are used at present to generate electricity and in combined heat and power systems. The gaseous biofuel methane is produced by the anaerobic digestion of sewage, in landfills and is also used for electricity generation and for heat and power systems. It is the liquid biofuels that will be used to replace the fossil fuels petrol and diesel, and they have attracted much attention. At present liquid biofuels can be divided into first-, second- and third-generation biofuels. The first-generation biofuels consist of ethanol which is used to either supplement or replace petrol and biodiesel, as a replacement for diesel. Ethanol is produced from either sugar or starch which is extracted from crops such as sugarcane, sugarbeet, wheat and maize and can save 30–80% of greenhouse gas emissions when compared with petrol. Biodiesel is produced from plant oils and animal fats and can save between 44 and 70% greenhouse gases compared with diesel. At first glance this situation appears ideal but there are problems with first-generation biofuel production and supply. One problem is the amount of energy that is used to produce and convert the crops into biofuels. Another problem is the amount of biofuels needed to replace fossil transport fuels. In the UK, in 2006 19,918 million t of petrol and 23,989 million t of diesel were used, that is a total of 43,907 million t. To supply this tonnage is a formidable task. For example, to supply the diesel required in the UK using the oil-seed crop rapeseed, 113% of the agricultural land would be needed. This is clearly not possible and even at modest levels of diesel replacement the biofuel crops would compete with food crops. It is this feature that has brought forward many objections to biofuels, and they have been blamed for some food shortages; however, in reality food prices are influenced by a number of factors. Converting sensitive lands such as rainforests to grow biofuel crops has also engendered justifiable resistance. In addition, crops such as wheat and other starch-containing crops require considerable processing and energy input, and when investigated by life-cycle analysis show only marginal gains in energy.

However, the resistance to biofuels need not be the case as the first-generation biofuels bioethanol and biodiesel were only intended to be used as a 5% addition to fossil fuels to comply with the EU directives, and to show that fossil fuels could be replaced. It was clear that any more would compromise food crops. It is the second- and third-generation biofuels that should replace the bulk of the transport fossil fuels. The second-generation biofuels are ethanol, produced directly from lignocellulose, and the gasification of lignocellulose and waste organic materials producing petrol, diesel, methanol and dimethyl ether. Lignocellulose and organic wastes are available in large quantities and their use does not compromise food crops. Lignocellulose is often the discarded portion of food crops such as straw. The third-generation biofuels are hydrogen, produced either by the gasification of lignocellulose or directly by microalgae, and biodiesel produced from oil accumulated by microalgae. These second- and third-generation biofuels should not compromise food crops, but to bring these fuels into production will require both research and investment. To stop the use of land areas, such as the rainforests, for first-generation biofuels will probably require legislation.

Acronyms, Abbreviations and Units

ACC	acetyl CoA carboxylase
AGFB	atmospheric gasifier fluidized bed
AFC	alkaline fuel cell
ATP	adenosine triphosphate
Bar	measurement of pressure = 100,000 pascal
bbl	barrel of oil = 159 litres
BERR	Business Enterprise and Regulatory Reform Department
BHT	butylated hydroxytoluene
BIGCC	biomass integrated gasifier combined cycle
BRI	Bioengineering Resources Inc.
BSFC	brake specific fuel consumption
BTDC	before top dead centre
Bt	*Bacillus thuringiensis*
BTL	biomass to liquid
Btu	British thermal unit
BOCLE	ball on cylinder lubrication evaluator
CA	crankshaft angle
CAP	common agricultural policy (EU)
CBG	compressed biogas
CCGT	combined cycle gas turbine
CC & S	carbon dioxide capture and storage
CDM	clean development mechanism
CFC	chlorofluorocarbons
CHP	combined heat and power
CNG	compressed natural gas
COP	Conference of the Parties
cP	centipoise (measure of viscosity)
CRGT	chemically recuperated gas turbine
cSt	centistoke (measure of viscosity)
Defra	Department of Food and Agriculture
DC	direct current
DDGS	distiller's dried grain with solubles
DICI	direct injection compression ignition
DISI	direct injection spark ignition
DME	dimethyl ether
DMFC	direct methanol fuel cell
DNA	deoxyribonucleic acid
Dti	Department of Trade and Industry

EBB	European Biodiesel Board
EGR	exhaust gases recycle
EIA	Energy Information Administration
EJ	exajoules (10^{18})
EOR	enhanced oil recovery
EPA	Environmental Protection Agency
ETBE	ethyl tertiary butyl ether
EU	European Union
EU ETS	European Union Emissions Trading Scheme
EU 25	European Union consisting of 25 countries
E85	petrol containing 85% ethanol
E95	petrol containing 95% ethanol
FAME	fatty acid methyl esters
FT	Fischer–Tropsch process
FT diesel	diesel produced by the Fischer–Tropsch process
FTIR	Fourier transformed infrared
FT petrol	petrol produced by the Fischer–Tropsch process
GHG	greenhouse gases
GLC	gas liquid chromatography
GTL	gas to liquid
Gtoe	gigatonne oil equivalent
ha	hectares = 10,000 m^2
HC	hydrocarbons
HFCS	high fructose corn syrup
HFRR	high frequency reciprocating rig
HGCA	Home Grown Cereals Association
ICE	internal combustion engine
IEA	International Energy Agency
IEE	Institute of Electrical Engineers
IFPRI	International Food Policy Research Institute
IGCC	integrated gasification and combined cycle
IOR	increased oil recovery
IPCC	International Panel on Climate Change
J	joules
JRC	Joint Research Centre
kW	kilowatt
kWh	kilowatt hour
kWht	kilowatt hour heat
LCA	life-cycle analysis
LHV	lower heating value (heat of condensation of water not included)
l	litre
LNG	liquefied natural gas
LPG	liquefied petroleum gases
LTFT	low temperature filterability
LULUCF	land use, land use change and forestry
MCFC	molten carbonate fuel cell
Mha	megahectare

MJ	megajoules
MPa	megapascal (unit of pressure; 1 MPa = 10 bar)
Mt	megatonnes
MTBE	methyl tertiary butyl ether
Mtoe	megatonnes oil equivalent at a LHV of 42 GJ/t
MWth	megawatt, thermal
M85	petrol containing 85% methanol
N	newton
NAD	nicotinamide adenine dinucleotide
NADP	nicotinamide adenine dinucleotide phosphate
NEV	net energy value
NG	natural gas
NIMBY	not in my back yard
NO_x	nitrous oxides
NREL	National Renewable Energy Laboratory
NSCA	National Society for Clean Air
OECD	Organisation for Economic Cooperation and Development
OPEC	Organisation of Petroleum Exporting Countries
PAFC	phosphoric acid fuel cell
PAR	photosynthetically active radiation
PEMFC	proton exchange membrane fuel cell
PEP	phosphoenol pyruvate
PgC	petagrams carbon (10^{15})
PHA	polyhydroxy alkanoate
PHB	polyhydroxy butyrate
PISI	port injection spark ignition
PM	particulate matter
ppbv	parts per billion by volume
ppmv	parts per million by volume
pptv	parts per trillion by volume
psi	pounds per square inch
PSI	photosystem I
PSII	photosystem II
PYRO	pyrogalol
RFS	renewable fuel standard
RME	rapeseed methyl ester
RNA	ribonucleic acid
ROC	renewable obligation certificate
RTFO	renewable transport fuel obligation
rubisco	ribulose bisphosphate carboxylase/oxygenase
SHF	separate hydrolysis and fermentation
SME	sunflower methyl ester
SOFC	solid oxide fuel cell
SPFC	solid polymer fuel cell
SRC	short rotation coppice
SSCF	simultaneous saccharification and co-fermentation
SSF	simultaneous saccharification and fermentation
Syngas	a mixture of carbon monoxide and hydrogen

TBHQ	*tert*-butyl hydroxyquinone
tcm	tera cubic metres (10^{12})
THF	tetrahydrofolate
TPES	total primary energy supply
TTW	tank-to-wheel
TWh	terawatt hours
t	tonnes
UNCED	United Nations Conference on Environment and Development
UNEP	United Nations Environment Programme
UNFCCC	United Nations Framework Convention on Climate Change
USDA	United States Department of Agriculture
WTT	well-to-tank
WTW	well-to-wheel
ZEV	zero emission vehicle

Conversions

Energy

1 Btu = 1055 J
1 calorie = 4.186 J
1 Mtoe = 42 PJ (10^{15} J)
1 Mtoe coal ~ 28 PJ
1 Mtoe oil ~ 42 PJ
1 kWh = 3.6 MJ
1 TWh = 3.6 PJ
1 GWh = 3.6 TJ
1 GWh = 0.000086 Mtoe
1 tonne (toe) = 41.868 GJ
1 average power station ~ 1 GW
3 tonnes dry wood (4.2 m^3) = 12,000 kWh

Pressure

1 pascal = 1 N/m^2
1 pascal = 1.02×10^{-5} atmospheres
1 pascal = 145×10^{-6} psi
1 MPa = 10 bar
1 bar = 100,000 pascals
1 bar = 14.5 psi
1 psi = 6894 pascals

Area/volume

1 ha = 10,000 m^2
1 barrel = 159 litres

Light

Irradiance (PAR) measured in Wm^{-2} or photons moles/m^2/s
$1W = 1$ joule s^{-1}
1 mole $= 6.02 \times 10^{23}$ photons (Avogadro's number)

Units

Prefix	Symbol	Factor
Exa	E	10^{18}
Peta	P	10^{15}
Tera	T	10^{12}
Giga	G	10^{9}
Mega	M	10^{6}
Kilo	k	10^{3}
Hecto	h	10^{2}
Deka	da	10
Deci	d	10^{-1}
Centi	c	10^{-2}
Milli	m	10^{-3}
Micro	μ	10^{-6}
Nano	n	10^{-9}
Pico	p	10^{-12}
Femto	f	10^{-15}
Atto	a	10^{-18}

1   Energy and Fossil Fuel Use

Introduction

At present we are living in a situation where the world's demand for energy continues to increase at a predicted annual rate of 1.8%, especially as countries develop, while at the same time the supply of energy appears limited. The reason for this is that 75–85% of the world's energy is supplied by the fossil fuels – coal, gas and oil (IEA, 2002; Quadrelli and Peterson, 2007) – and the supply of these is finite. In addition, the burning of fossil fuels has increased the atmospheric concentration of some greenhouse gases that are responsible for global warming. Other consequences of burning fossil fuels include the production of acid rain, smog and an increase in atmospheric particles. In addition, the world's population is expected to expand at about 1% per year, which will mean that global energy requirements will continue to rise. It is predicted that fossil fuels will continue to dominate the energy market for some time and oil will be the most heavily traded fuel. The Middle East contains the bulk of the oil reserves and, therefore, much of the global oil supply will increasingly be obtained from this area. This will increase the world's vulnerability to price shocks caused by oil supply disruption from this somewhat unstable area. Against this background, all countries' (including the UK's) access to adequate energy supplies will become increasingly important at a time when oil supplies are declining, such as the North Sea's oil and gas. Alternative sources of energy, which are renewable and with sustainable supplies, are required. Renewable energy sources can provide a constant supply of energy, and examples are hydroelectricity, wind and wave power, and geothermal- and biological-based fuels. It would be foolish to think that any one of these renewable energy sources could completely replace fossil fuels. However, if each of the renewable sources can make a contribution, when combined they may be able to replace fossil fuels, although this would probably need to be in conjunction with a reduction in energy use, and an increase in its efficiency. The challenge for all countries is therefore to move to a more secure, low-carbon energy production, without undermining their economic and social development.

In this book, the problems associated with fossil fuel use are outlined, and how the adoption of alternative fuels can mitigate global warming. Other chapters cover biologically produced, solid, gaseous and liquid fuel production with their advantages and disadvantages.

Fossil Fuel Use

In the past, the world's energy supply was based on wood, a renewable resource, which was used for cooking, heating and smelting. Later, water and wind power were harnessed and used especially throughout Europe. The Industrial Revolution was initiated using water power but this was soon replaced by coal, which had high energy content and was freely available. Oil and gas use developed after coal and in the face of what appeared to be unlimited supplies of fossil fuels, water and wind power were abandoned. The worldwide economic growth since the mid-20th century has been sustained by an increasing supply of fossil fuel. Huge infrastructures have been organized to supply these fuels, and whole communities have grown to extract fossil fuels, particularly coal.

The global use of energy, and therefore fossil fuels, has increased steadily ever since the Industrial Revolution in the 1800s, and Fig. 1.1 shows the world's total energy consumption in gigatonnes of oil equivalent (Gtoe, 10^9 t of oil equivalent) up to the present and predictions up to 2030. At present the annual consumption is around 10 Gtoe and the world's energy needs are projected to grow by 55% from 2005 to 2030 at an average rate of 1.8% (IEA, 2005a, 2007). Estimates of the current world energy demand in exajoules (10×10^{18} J) are given in Table 1.1 with a consensus value of 410 EJ. The increases predicted in the world's primary energy demand up to 2030 can be seen in Fig. 1.1 (IEA, 2005a). The predicted values may be underestimated because of the recent increases in the Indian and Chinese economies. These emerging economies are growing rapidly and China is increasing coal extraction and continues to build coal-fired power stations. The increases in oil consumption by Asia compared to North America and Europe can be seen in Fig. 1.2.

In the use of energy there is a correlation between average income and energy consumed. Figure 1.3 shows the relationship between average income and oil consumption. It is clear that as countries become more developed, the demand for energy will increase considerably, and this is happening in particular with China and India.

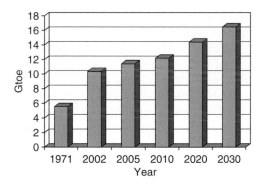

Fig. 1.1. Present and predicted world energy consumption (in Gtoe). (From IEA, 2005a, 2007.)

Table 1.1. Estimates of world energy demand in exajoules (10^{18} J).

Energy demand EJ (10^{18})	Reference
410.3	Hein (2005)
410.0	Bode (2006)
379.0	Reijnders (2006)
410.0	Odell (1999)
512.0	IEA (2005a)

The consensus is 410 EJ which is equivalent to 9.76 Gtoe.

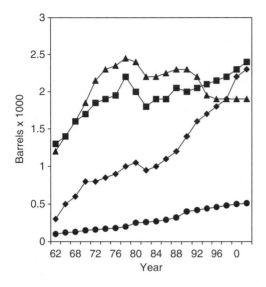

Fig. 1.2. Daily oil consumption (in 1000 barrels of 159l each). ▲ North America; ■ Europe; ◆ Asia-Pacific; ● Middle East. (Redrawn from Guseo *et al.*, 2007.)

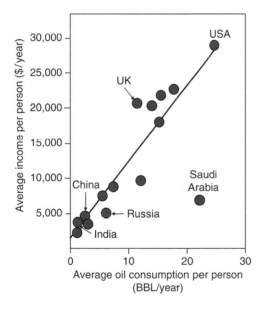

Fig. 1.3. The relationship between average income and oil consumption (1 barrel = 159l). (Redrawn from Alklett, 2005.)

As well as an increase in the combustion of fossil fuel, the pattern of consumption has changed dramatically. Coal fuelled the Industrial Revolution in the 1800s, and even by the 1930s, over 70% of the world's energy was still derived from coal. Since the mid-2000s, oil and gas have replaced coal as the main world energy sources, with smaller contributions made by nuclear, biomass and hydroelectric resources. Figure 1.4 shows the sources of world primary energy supply represented as percentages. It is clear that the supply is now dominated by gas and oil.

Table 1.2 gives the current and predicted sources of primary energy in the world in terms of Gtoe. Nuclear power, once thought to be a limitless source of energy, has increased slowly since the late 1980s due to problems of radioactive waste disposal

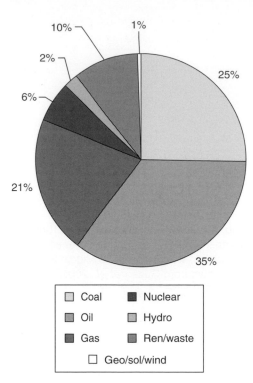

Fig. 1.4. Present sources of world primary energy supply for 2005 (shown as percentages of a total of 11.43 Gtoe). (Redrawn from IEA, 2007.)

Table 1.2. Current and predicted sources of world primary energy demand (in Gtoe) for a number of energy sources. (Adapted from IEA, 2005a.)

	1971	2002	2010	2020	2030
Coal	1.41	2.39	2.76	3.19	3.60
Oil	2.52	3.92	4.46	5.23	5.93
Gas	0.89	2.19	2.70	3.45	4.13
Nuclear	0.029	0.69	0.78	0.78	0.76
Hydro	0.10	0.22	0.28	0.32	0.37
Biomass	0.49	0.76	0.83	0.88	0.92
Renewables	0.004	0.06	0.10	0.16	0.26
Total	5.54	10.35	12.19	14.40	16.49

and decommissioning of old power stations. The nuclear contribution has remained stable, whereas biomass and renewables have shown a steady increase. In 2004, fossil fuels were used to produce 72% of the world's electricity with coal producing 39%, oil 8% and gas 25% (Quadrelli and Peterson, 2007).

The use of gas is predicted to continue to replace coal for electricity generation as it is a cleaner fuel producing fewer greenhouse gases. Coal is predicted to increase by 50%, whereas gas is expected to increase by 88%. The reduction in carbon dioxide when switching from coal to gas follows the formulae given below:

$$(\text{coal}) \ 2(CH) + 2\tfrac{1}{2}O_2 = 2CO_2 + H_2O \tag{1.1}$$

$$(\text{natural gas}) \ CH_4 + 2O_2 = CO_2 + 2H_2O \tag{1.2}$$

The pattern of change in energy use in the UK has mirrored the global pattern, where coal has been superseded by gas for electricity generation (Fig. 1.5), and the nuclear and the renewables sectors have also increased their contribution. In Fig. 1.5, nuclear power is combined with renewable energy sources but as the supply of uranium is finite, some regard nuclear power as non-renewable.

The overall fuel consumption by the various domestic and industrial sectors is given in Table 1.3, which shows that transport uses 37.1% of the energy. In the case of electricity generation, large losses of energy occur during generation (55.2 Mtoe) and distribution (19.1 Mtoe) from a total energy consumption of 232.1 Mtoe. Combined, these are 74.7 million t of oil equivalents (Mtoe) or 31.9% of the total energy produced. This is a consequence of large centralized electricity generation, where waste heat cannot be used, and the transportation of electricity over long distance, which involves losses.

An outline of the major flows of energy within the UK is shown in Fig. 1.6. The complete pattern of flow is more complex than that shown in the figure, which only shows the major routes, but it is clear that oil is used exclusively to produce transport fuels. Gas is used both for heat and electricity generation, whereas coal is mainly used for electricity generation.

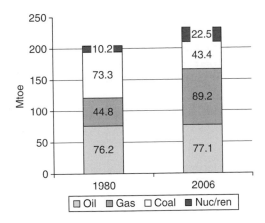

Fig. 1.5. UK energy consumption changes since 1980. (From BERR, 2007.)

Table 1.3. Overall fuel consumption by industry in the UK for 2005 (in Mtoe). (Adapted from Dti, 2006a.)

Fuel	Industry	Domestic	Transport	Services[a]	Total
Coal	2.0	0.7	–	–	2.7
Gas	12.8	32.8	–	9.2	54.8
Oil	7.1	3.1	58.5	1.8	70.4
Electricity	10.2	10.0	0.7	8.7	29.7
Renewables	1.0	0.3	–	0.6	1.9
Total	32.1	47.0	59.2	20.0	159.5
	(20.1%)	(29.5%)	(37.1%)	(12.5%)	

[a]Includes agriculture.
( ), percentage of total.

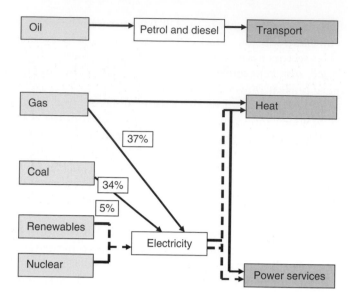

Fig. 1.6. The major flows of energy in the UK. (Modified from Dti, 2006a.)

The fuels used for electricity generation in the world and the UK are given in Table 1.4. It is clear that worldwide coal is still the major fuel in electricity generation, with gas, nuclear and hydroelectric resources making similar contributions. The renewable electricity generation such as wind, solar and biomass only contributed 2.2% of the total worldwide. However, in the EU and UK, gas is used to produce a large proportion of the area's electricity. The move from coal to gas for electricity generation in the UK can be seen in Fig. 1.7, and this has altered the overall fuel usage (Fig. 1.5).

Table 1.4. Fuels used to generate electricity in 2005 for the world, EU (25) and the UK in terawatt hours (TWh). (From Dti, 2006a; BERR, 2007; IEA, 2007.)

Fuel	World	%	EU (25)	%	UK	%
Coal	7350	40.2	1000	30.2	136.6	34.15
Oil	1200	6.5	138	4.2	5.4	1.35
Gas	3596	19.6	663	20.0	153.2	38.3
Biomass	161	0.88	57	1.7	8.1	2.0
Waste	64	0.35	27	0.82	4.81	1.2
Nuclear	2768	15.1	997	30.1	81.6	20.4
Hydro	299	1.6	340	10.3	7.89	1.97
Geothermal	57	0.31	5.4	0.16	0	–
Solar PV	1.6	0.009	1.5	0.045	8	2.0
Solar heat	1.1	0.006	0	–	0	–
Wind	101	0.006	70.5	2.13	2.91	0.73
Tide	0.56	0.003	0.53	0.016	0	–
Other sources[a]	8.8	0.048	7.0	0.21	0	–
Total	18,306		3,311		400.5	

[a]The figure includes imports.

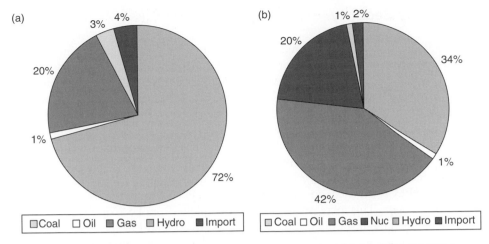

Fig. 1.7. Changes in fuel used for electricity generation in the UK. (a) 1990; (b) 2004. (Modified from Cockroft and Kelly, 2006.)

The contribution of renewable energy sources to electricity generation is very small for the UK, so it does not show in Fig. 1.7, but a more detailed list of renewable sources of energy in the UK in 1990 and 2006 is given in Table 1.5. The total renewable energy was 4.4 Mtoe in 2006 from a total energy use of 159.5 Mtoe.

The table excludes nuclear power, but the two sources that have increased are landfill gas (methane) and wind/wave energy. Landfill sites used to be disposal sites and no attempt was made to collect the methane gas formed in the anaerobic digestion of the organic components of the waste. This has changed now and collection systems are installed during the filling of the landfill site. Wind power technology is now fairly mature and wind farms have been constructed in areas of consistently high wind. The positioning of some wind farms has seen objections raised due to noise, effects on birds and interference with radar. Many of these objections may not be found with offshore wind farms. Wave power is still under development with a number of systems designed to extract energy from waves directly from ocean currents. The row labelled 'Other biofuels' includes ethanol and biodiesel, which have seen a rapid increase in production in the last few years.

Table 1.5. Renewable energy sources in the UK from 1990 to 2006 (in 1000 t of oil equivalent). (Adapted from BERR, 2007.)

Source	1990	2000	2004	2005	2006
Geothermal/solar	7.2	12.0	25.7	30.9	37.8
Wind and wave	0.8	81.3	166.4	249.7	363.3
Hydro	447.7	437.3	416.5	423.2	395.9
Landfill gas	79.8	731.2	1326.7	1420.8	1464.7
Sewage gas	138.2	168.7	176.6	179.1	200.3
Wood	174.1	425.0	399.8	285.1	285.1
Waste combustion	100.8	374.8	463.2	460.0	512.7
Other biofuels	71.9	265.0	710.1	1,191.3	1,170.4
Total	1020.5	2495.2	3685.1	4240.0	4430.1

Fuel Supply Security

In the 1950s, coal remained the main source of fuel for home heating, industry and electricity generation. Nuclear power started in the 1960s but supplied only 3.5% of western Europe's electricity by 1970. As the economies of European countries increased, oil was increasingly imported and much of it from the Middle East. In 1971, the USA became for the first time an oil importer, which increased worldwide oil demand, and at the same time Kuwait and Libya reduced oil production. These developments were the first indication of the dependence of developed countries on imported fossil fuels, especially oil, and the need for secure supplies of fuel. The first crisis in oil supply was caused by the Israel–Arab War in 1973 when the Organisation of Petroleum Exporting Countries (OPEC) imposed an oil embargo on the USA and the Netherlands and reduced production. This increased the oil prices rapidly (from US$3 to US$11 a barrel), which highlighted the instability of oil supplies in the UK and Europe. This encouraged non-OPEC countries to search for oil. In 1967, the UK started piping natural gas ashore from the North Sea and later on in the 1970s and 1980s, oil was discovered in the deeper parts of the North Sea. The supply of natural gas encouraged the switch from 'town gas' produced from coal, to natural gas for home heating and cooking. In 1981, the UK was self-sufficient in oil as the North Sea oil fields were exploited (Hammond, 1998). The Iran–Iraq war in 1979–1982 caused a second crisis in oil when the price rose to US$38 a barrel. In the UK, in 1979, the energy sector was privatized with the exception of the nuclear sector where there were concerns about the costs of decommissioning of nuclear plants. The importation of cheap coal was also permitted and this, in combination with a cheap supply of natural gas, meant that deep coal mining was drastically reduced. The increased supply of natural gas saw the introduction of combined cycle gas turbine (CCGT) for electricity generation, as this system released less carbon dioxide. This 'dash for gas' also reduced the demand for coal, which dropped from 60% in 1972 to 18.7% in 2006. Continued instability in the oil and gas markets and the depletion of the North Sea oil and gas supplies have strengthened the need for fuel security in the UK. In 2008, there was another oil crisis where oil peaked at around US$150 a barrel, which emphasizes the dependence of developed countries on oil. The reasons for these rapid rises were unclear, as there was no war interrupting supplies. Oil supplies appear to be adequate, so it may be lack of refining capacity and speculation that caused the rises in price.

Fossil Fuel Reserves

Nobody would dispute that fossil fuels supplies are finite, but what is disputed is the extent of the reserves remaining, and how long these will last. Over the years, there have been a large number of estimates based on present consumption, reserves and predicted new sources (Grubb, 2001; Bentley, 2002; Greene et al., 2006).

New oil fields have been found both on land and under the sea bed, but these new fields are being found in increasingly hostile environments. It has been concluded that the world is halfway through its recoverable oil, except for the Middle East (Bentley, 2002). Table 1.6 shows some of the estimates for the peak of production, known as 'peak oil' (Fig. 1.8), a time after which production declines, and the time when fossil fuels run out.

The International Energy Agency (IEA, 2005a) has predicted that the supply of crude oil will peak around 2014 and then decline and coal will last until 2200 (Evans,

Table 1.6. Estimated life of fossil fuels.

Author	Peak date	Run-out date
Fulkerson *et al.* (1990)		Gas 2047
		Coal 2180
Odell (1999)	Oil 2030–2060	2120
	Gas 2050–2090	
Evans (2000)	Oil 2014	Gas 2080
	Gas 2020	Coal 2200
Grubb (2001)	Oil 2010–2020	
Laherrere (2001)	Oil 2020	
Bentley (2002)	Oil 2020	
	Gas 2020	
BP (2005)		Oil 2050
Greene *et al.* (2006)	Oil 2020	
Crookes (2006)		Oil 2043

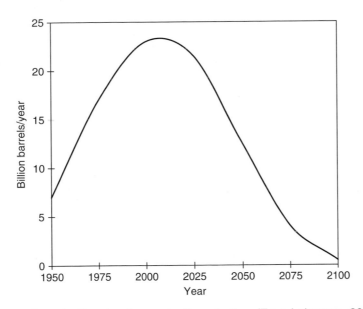

Fig. 1.8. Future oil production showing peak oil production. (From Laherrere, 2001.)

1999). The decline in available coal and crude oil should cause the prices of these fuels to rise, which would limit their use.

Despite the general agreement in these dates, there is still considerable debate over the quantity of known and unknown oil reserves. These figures will clearly have a considerable influence on the lifetime of the oil production. The estimates given in Table 1.6 are based on conventional oil reserves. Data indicate that two-thirds of oil-producing countries are past their peak of conventional oil production, including the USA, Iran, Libya, Indonesia, the UK and Norway (Bentley *et al.*, 2007). The estimates have taken into account the proved oil, probable reserves of oil and the rate of discovery. The rate of discovery of new oil fields controls oil production. When oil production and oil discovery are plotted for the UK, it can be seen that discoveries

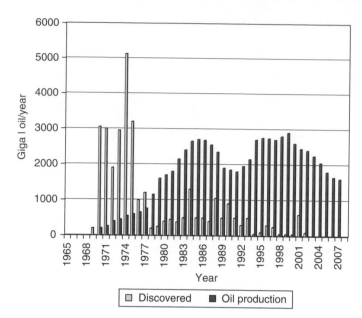

Fig. 1.9. The discovery of proved and probable oil (2P) and oil production for the UK. Oil discovery measured as millions of barrels of oil per year and production thousands of barrels per day. (From Bentley *et al.*, 2007.)

peaked in the 1970s, while production peaked in 1999 (Fig. 1.9). In mitigation the UK has a few potential sources of oil, including the deep Atlantic, on land, and small pockets of oil in existing fields. Other estimates suggest that 64 of the 98 oil-producing countries have passed their peak of oil production so that the same conditions apply to many other countries (see www.lastoilshock.com). Economic factors will also affect oil availability as higher prices will encourage exploration, expensive recovery, the use of marginal oil fields and depress oil demand, though these effects are unlikely to greatly affect estimates of oil reserves (Bentley *et al.*, 2007).

However, there are technologies available that can be applied to extract more oil from existing oil fields and there are also unconventional oil sources. Improved oil recovery (IOR) involves techniques like horizontal drilling and improved management. Enhanced oil recovery (EOR) involves technologies to mobilize oil trapped in the well and includes gas injection, steam flooding, polymer addition and combustion *in situ*. Depending on the geology of the oil field, the oil enhancement can range from 10 to 100%. Recent studies have indicated that EOR can temporarily increase the rate of oil production, but the consequence is an increase in the rate of depletion (Gowdy and Julia, 2007). Figure 1.10 shows the oil production from the Forties oil-field where EOR (carbon dioxide flood) was applied in 1987.

Taken in the context of the history of mankind, the use of fossil fuels has been with us for only a short time, as can be seen in Fig. 1.11 (Aleklett, 2005). This means that the stocks have to be conserved, and alternatives introduced that are not reliant on plants and animals that died some 1–100 million years ago. A comparison between biomass currently grown and fossil fuel production in the industrial carbon cycle is shown in Fig. 1.12. It is clear that fossil fuels are not being replaced as conditions are different now and the timescales preclude any form of replacement.

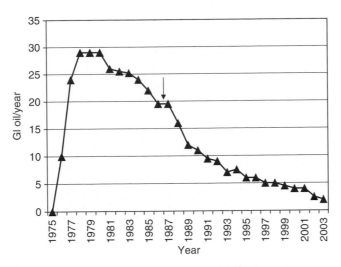

Fig. 1.10. Oil production in the North Sea Forties oilfield (↓) through carbon dioxide flood. (From Gowdy and Julia, 2007.)

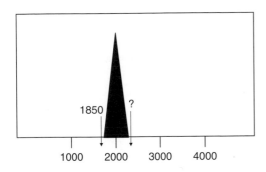

Fig. 1.11. The short history of fossil fuels. (From Aleklett, 2005.)

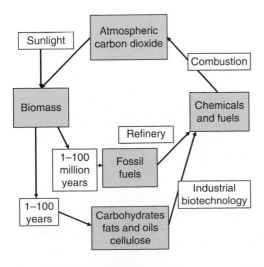

Fig. 1.12. Petrochemical carbon cycle compared with the use of biomass. (Modified from Faaij, 2006.)

Three non-conventional sources of oil exist: heavy oil, oil shales and tar sands. Reserves of heavy oils are found in Venezuela and the oil is mobile under normal–well conditions but is extremely viscous on extraction. The bulk of oil shales are found in the USA with some in Estonia, Brazil and China. Shale oil is unfinished oil made up of kerogen oil because it has not been exposed to high temperatures. Shales deposits can contain between 4 and 40% kerogen, which can be released when the rock is heated to 300–400°C.

Tar sands contain 10–15% bitumen and are found close to the surface, so that it can be recovered using open cast techniques. The mined and in situ treatment involve heating to allow the oil to separate from the sand. Once extracted, the tar is heated to 500°C to yield kerosene and other distillates. Although tar sands do occur world-wide, 85% of the tar sands are found in Alberta, Canada. Both shale oil and tar sands require considerable energy to extract and process and therefore produce more carbon dioxide. These sources have been of little economic interest until recently.

The extent of the known reserves and those reserves to be discovered have been subject to a large number of estimates, and examples of two are given in Tables 1.7–1.9. There is reasonable agreement between the two estimates where it is clear that the bulk of the fossil fuel reserves are in coal. At present just below half of the total conventional oil reserves have been consumed, although there are considerable unconventional reserves. The conventional oil reserves and those predicted constitute

Table 1.7. Carbon content of global fossil fuels in Gt carbon (10^9 t). (From IPCC, 1996.)

	Reserves	Reserves to be discovered 50% probability	Reserves that require technological advances to be extracted	Total Gt C
Oil				
Conventional	120	50	–	170
Unconventional	142	–	180	322
Gas				
Conventional	73.4	67.3	–	140.8
Unconventional	105.6	–	306	411.6
Coal	650.2	–	2587.7	3237.9
Total	1091.2	117.3		4282.2

Table 1.8. Fossil fuel consumption and reserves in Gt carbon (10^9 t). (From Grubb, 2001; UNEP, 2000.)

	Consumption 1860–1998	Consumption 1998	Reserves	Reserves to be discovered	Total Gt C
Oil					
Conventional	97	2.65	120	121	241
Unconventional	6	0.18	102	305	407
Gas					
Conventional	36	1.23	83	170	253
Unconventional	1	0.06	144	364	509
Coal	155	2.40	533	4618	5151
Total	294	6.53	983	5579	6562

Table 1.9. Oil and gas reserves in various regions in 10^9 t. (From BP Statistical Review, 2005; Cedigaz 2004.)

Region	2002	2004	Gas 2004 (tcm)[a]
North America	6.8	8.3	7.5
South and Central America	13.5	13.8	7.3
Europe and Eurasia	13.3	19.0	63.3
Middle East	93.5	100.1	71.6
Africa	10.6	15.3	13.8
Asia Pacific	5.3	5.6	16.3
Total	142.9	162.1	180

[a]tcm: tera(10^{12}) cubic metres.

170 and 139 gigatonnes (Gt) of carbon, respectively. With oil containing 86% carbon, this represents 197.6 and 280.2 Gt of oil, respectively. These figures correlate well with the estimates of oil and gas reserves given on a regional basis in Table 1.9. The consumption rates have been given as 3.5 and 3.08 Gt per annum, respectively, which, if correct, means that the reserves will last between 40 and 53 years.

Regional reserves

In addition to having a finite life, fossil fuel reserves, in particular oil, are not evenly distributed. Table 1.9 gives various regions where it is clear that the major reserves of oil are found, with the majority in the Middle East (61.7%), and gas is split between the Middle East and Europe and Eurasia. This distribution will have considerable effects on supplies at political and economic levels and make it imperative that countries secure their energy supplies in the future.

Methane hydrates

Another potential future energy source is the methane hydrates. Methane hydrates are methane molecules encaged in a lattice of water molecules with a crystal structure of $(CH_4)_5(H_2O)_{75}$. Low temperature and high pressure induce hydrate formation, which when dissociated can release 164 times their own volume of methane. The amount of methane contained in the hydrates has been estimated at 21×10^{15} m^3, equivalent to 11,000 Gt carbon (Kvenvolden, 1999). There are onshore (permafrost) and offshore (below 2000 m) deep sea hydrate deposits, and several countries have projects to exploit these (Lee and Holder, 2001; Glasby, 2003). Possible methods of exploitation are heating, depressurization and inhibitor injection to dissociate the hydrate. However, these methods do have disadvantages due to the instability of the hydrates, collection of the gas, instability of deep sea sediments and uncontrolled release of methane into the atmosphere.

UK supply of fossil fuels

For some time the UK has been self-sufficient in supplies of oil and gas, but the supplies from the North Sea are declining and it is predicted that future supplies will have to be imported. Figure 1.13 shows the predicted decline in oil and gas production from the North Sea, and Fig. 1.14 shows the predicted gas imports that will be needed by 2020.

It is clear that the UK will increasingly have to import liquid fuels. As the bulk of the reserves of oil are in the Middle East and gas reserves in Siberia, the imports will be coming from unstable areas where interruptions may occur at any time. In these conditions, any UK production of energy or fuel will go some way to secure the supply of that energy.

Sustainable Fuel Sources

Renewable energy means an energy source that can be continually replaced, such as solar energy and plant materials, where the energy is obtained from the sun during

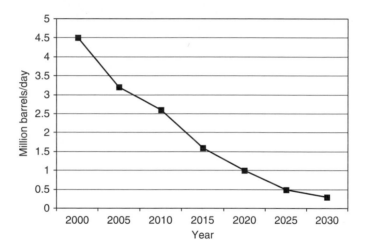

Fig. 1.13. Predictions on the UK oil production. (From Dti, 2006b.)

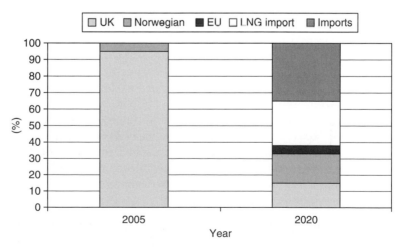

Fig. 1.14. Predicted changes in UK produced and imports of liquid natural gas (LNG). (From Dti, 2006a.)

photosynthesis. However, to allow indefinite use, renewable sources should not be depleted faster than the source can renew itself.

There have been a number of definitions for sustainability. One definition was 'development that meets the needs of the present without compromising the ability of future generations to meet their needs' (Glasby, 2003). It has also been defined as 'to prolong the productive use of our natural resources over time, while at the same time retaining the integrity of their bases, thereby enabling their continuity' (de Paula and Cavalcanti, 2000).

Sustainable development focuses on the long term, using scientific developments to allow a switch from the use of finite resources to those which can be renewed. Sustainability has also become a political movement involving groups working to save the environment.

Another term used for non-fossil energy sources is 'carbon-neutral', which means that either the energy production yields no carbon dioxide, such as solar and nuclear power, or the process only releases carbon dioxide previously fixed in photosynthesis (Fig. 1.15). In determining the carbon dioxide reduction for renewable energy sources, life-cycle analysis will determine the fossil fuel input into the production of the fuel and carbon dioxide produced. These points must be taken into consideration when the carbon dioxide savings are determined, and when applied to biofuels many are less than 100% carbon-neutral.

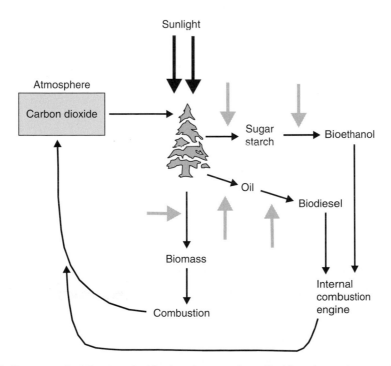

Fig. 1.15. Concept of carbon-neutral fuels, where carbon dioxide released on combustion has been previously fixed in photosynthesis. However, the arrows indicate that energy, probably from fossil fuels, has been expended in harvesting, extraction and processing of these fuels. This will reduce the amounts of carbon dioxide saved.

Conclusions

It is clear that the world's energy demand will continue to increase in developed countries and more particularly in developing countries such as China and India. The pattern of fossil fuel use is also changing with coal being replaced with gas for electricity generation. At the same time, renewable sources of energy are being developed, in particular biogas and wind power. It is clear that the supply of fossil fuels is finite, considering how it was produced, but the discussion centres around how long the stocks will last and the extent of the fossil fuel reserves. The world's dependence on a constant supply of energy means that whatever the estimate of the fossil fuel reserves, renewable sources need to be introduced as rapidly as possible.

2   Consequences of Burning Fossil Fuel

Introduction

Weather and climate have profound effects on all forms of life on Earth, and the major influence on climate is the energy derived from sunlight. Weather has been defined as 'the fluctuating state of the atmosphere in terms of temperature, wind, precipitation (rain, hail, and snow), clouds, fog and other conditions' (IPCC, 1996). Climate refers to the mean values of the weather and its variation over time and position. Perhaps in simple terms, weather is the conditions experienced by the individual and climate is the conditions experienced by countries and other land areas. The atmospheric circulation and its interaction with large-scale ocean currents, land masses and the sun determine climate. The effect on climate is known as forcing, and the most important determinant is the sun.

The atmosphere is composed mainly of nitrogen (78%) and oxygen (20.95%) and has recognizable layers starting at the surface with the troposphere, followed by the stratosphere, mesosphere and thermosphere. The troposphere is the layer in which weather occurs and contains 90% of the gases that make up the atmosphere. In this layer there are a number of gases other than nitrogen and oxygen, but only present in trace amounts, which include carbon dioxide, methane, nitrous oxide and ozone. In addition to these gases, the atmosphere contains solid and liquid particles and clouds (water vapour). Ozone in the atmosphere is found in trace amounts at all levels but is at a maximum (8–10 ppm) in the stratosphere, known as the ozone layer. All these gases absorb and emit infrared radiation and are collectively known as the greenhouse gases (Table 2.1).

All the gases, except the chlorofluorocarbons (CFCs), can be formed in nature and the balance between their production and elimination ensures that the global temperature is constant and sufficient to maintain life on the planet. Water vapour, which is also a greenhouse gas, is the most variable and is not normally included with greenhouse gases.

About half the radiation which arrives from the sun is in the visible range (short wave, 400–700 nm) and the other half is made up of near infrared (1200–2500 nm) and ultraviolet (290–400 nm). The land surface, consisting of soil and vegetation, influences how much of the sunlight energy is adsorbed by the Earth's surface and how much is returned to the atmosphere. Ice in the form of glaciers, snowfields and sea ice reflect radiation, whereas dark surfaces adsorb radiation.

The Earth's surface temperature is considerably lower than the sun's (14°C compared with 3000°C) and as a consequence it radiates any energy as infrared (Fig. 2.1). This is known as black body radiation. Some of the energy adsorbed by the Earth's surface is lost by radiation into space. However, the loss of this infrared radiation is affected by the greenhouse gases. These gases absorb the radiation and direct some of it back to the Earth's surface. The outcome of this return of energy to the surface is

Table 2.1. Greenhouse gases[a] and their contribution to global warming (1980–1990). (Adapted from IPCC, 1996.)

Gas	Contribution to global warming (%)	Global warming potential compared to CO_2	Source
Carbon dioxide	55	1	Natural & mankind
Chlorofluorocarbons (CFCs)	24	24	Only mankind
Methane	15	21	Natural & mankind
Ozone	ND	ND	Natural
Nitrous oxide (N_2O)	6	310	Natural & mankind
Water vapour	ND	ND	Natural & mankind

ND: Not determined.
[a]Water vapour is also a greenhouse gas but its contribution is difficult to determine.

that the average temperature of the Earth's surface is 14°C rather than –19°C, which would be the case if all the radiated energy was lost into space (Figs 2.1 and 2.2). Some of the greenhouse gases are better at absorbing radiation than others and therefore their effects are not directly linked to their atmospheric concentrations (Table 2.1). For example, methane is 21 times as efficient as carbon dioxide in adsorbing infrared radiation, but as the concentration of methane is 350 times less than carbon dioxide its contribution to global warming is 15% compared with carbon dioxide at 55% (Table 2.1). If the greenhouse gas concentration remains constant, the energy

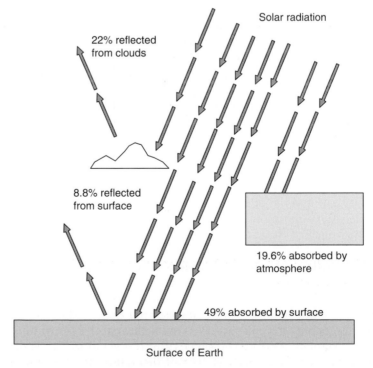

Solar radiation

22% reflected from clouds

8.8% reflected from surface

19.6% absorbed by atmosphere

49% absorbed by surface

Surface of Earth

Fig. 2.1. Energy in the form of radiation from the sun arriving at the Earth. (From Scragg, 2005.)

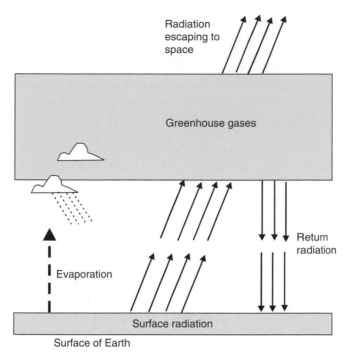

Chapter 2

Radiation escaping to space

Greenhouse gases

Return radiation

Evaporation

Surface radiation

Surface of Earth

Fig. 2.2. The radiation of long waves from the Earth's surface and the effect of greenhouse gases on this heat loss. (Scragg, 2005.)

Table 2.2. The effect of human activity on greenhouse gas levels.

Gas	Pre-industrial levels (1750–1800)[a]	Post-industrial levels (1990)[a]	(2000)[b]	(2005)[c]
Carbon dioxide (ppmv)	280	353	368	379
Methane (ppmv)	0.8	1.72	1.75	1.77
Nitrous oxide (ppbv)	288	310	316	319
Chlorofluorocarbons (pptv)	0	764	–	–

[a]Houghton *et al.* (1990).
[b]IPCC (1996).
[c]IPCC (2007).

input from the sun is balanced by the proportion lost by radiation and the global temperature remains stable.

The Effects of Industrial (Anthropogenic) Activity on Greenhouse Gases

The climate system has remained relatively stable for at least 1000 years prior to the Industrial Revolution as measured by ice-core determinations. The atmosphere has maintained a balance between carbon dioxide fixed by photosynthesis in vegetation and soil and the production of carbon dioxide from the decomposition of biological materials and plant and animal respiration. In this balance the oceans have acted as a large carbon dioxide sink, but the adsorption rate of carbon dioxide into the oceans is slow. Figure 2.3 shows the carbon dioxide flows between the oceans, forests, soil and atmosphere with the concentration held in these areas in gigatonnes (Gt – 10^9 t) of carbon dioxide.

Since the start of the Industrial Revolution, atmospheric concentrations of carbon dioxide, methane and nitrogen oxides have all increased. The increases in carbon dioxide, methane and nitrous oxides appear to be due to human activities, as shown in Table 2.2. The reasons for this increase are the burning of fossil fuels, deforestation, agricultural activities and the introduction of CFCs. The burning of fossil fuels oil, coal and gas releases carbon dioxide trapped by plants and animals millions of years ago (Fig. 2.3). The overall pattern of greenhouse gas production worldwide is shown in Fig. 2.4, where the major source of carbon dioxide is energy utilization. Carbon dioxide is also produced with land use change such as deforestation and cultivation. The major source of methane and nitrous oxide is agriculture.

The term 'global warming' is not new as it was first coined by J.B. Fourier in 1827, and in 1860 J.H. Tyndall measured the heat adsorbed by carbon dioxide.

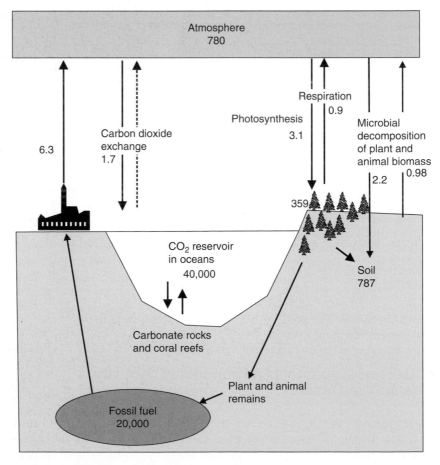

Fig. 2.3. Global carbon flow between the atmosphere, oceans, forests and the contribution made by fossil-fuel burning. The values are gigatonnes (Gt) of carbon. (From Kirschbaum, 2003.)

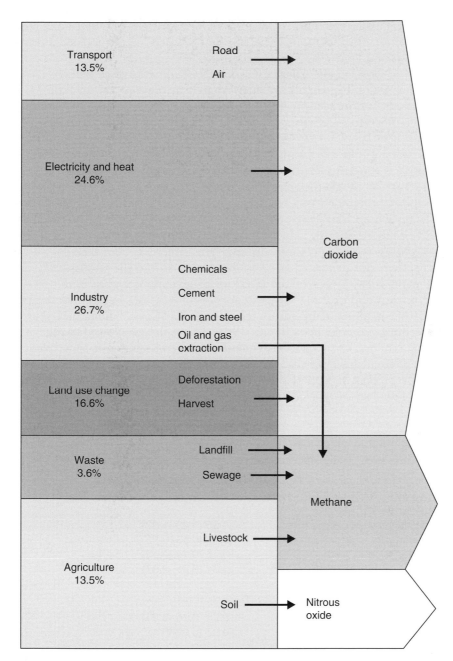

Fig. 2.4. World sources of greenhouse gases. (Redrawn from the World Resources Institute, 2006.)

However, it was not until 1938 that G. Callender showed for the first time that the world's temperature was increasing and, in 1957, that carbon dioxide could be the cause. In 1965, the US government first looked into the connection between atmospheric carbon dioxide increases and the burning of fossil fuels, which was confirmed at the 1979 World Climate Conference in Geneva.

A panel of experts was asked to study the effect of greenhouse gases on climate change by the United Nations. These invitations led to the introduction of the Intergovernmental Panel on Climate Change (IPCC) in 1988.

The IPCC in its report *Climate Change 2007: The Physical Science Basis* concluded from a number of observations that 'warming of the climate system is unequivocal, as is now evident from observations of increases in global average air and ocean temperatures, widespread melting of snow and ice, and rising global mean sea level'.

There are a number of factors that can influence global warming in addition to greenhouse gas concentrations. One of the methods used to determine the effects of various factors, including the greenhouse gases, is to calculate their radiative forcing. Radiative forcing is a measure of the influence that a factor has on the world's energy balance – both positive and negative (Table 2.3). The greenhouse gases have all positive radiative forcing values but aerosol and clouds reduce radiation reaching the surface, thus reducing global warming. This effect has been seen in cases where volcanic activity has deposited large amounts of dust into the atmosphere.

The main consequence of the build-up of greenhouse gases is the increase in global temperature (Fig. 2.5). The consequences of increases in greenhouse gases will be an increase in global temperature from 0.5 to 6.0°C, depending on the measures taken to reduce their emissions (IPCC, 1996).This is the most rapid change in global temperature for the last 10,000 years and will have a number of consequences and a number of scenarios put forward. A summary of the impact of climate change is given in Wuebbles *et al.* (1999) and Stern (2006), and this impact includes rise in sea level, loss of sea ice and glaciers, more extreme weather, and increases in desert areas

Table 2.3. Radiative forcing factors. (Modified from IPCC, 2007.)

Factor	Radiative forcing values W m^2
Greenhouse gases	
Carbon dioxide	1.66 (1.49–1.82)
Methane (CH_4)	0.48 (0.43–0.53)
Nitrous oxide (N_2O)	0.16 (0.14–0.18)
CFCs	0.34 (0.31–0.37)
Ozone	
Stratospheric	−0.05 (−0.15–0.05)
Tropospheric	0.35 (0.25–0.65)
Water vapour from methane	0.07 (0.02–0.12)
Surface albedo	
Land use	−0.2 (−0.4–0.0)
Black carbon on snow	0.1 (0.0–0.2)
Total aerosol	
Direct effect	−0.5 (−0.9–0.1)
Cloud albedo	−0.7 (−1.9–0.3)
Linear contrails	0.01 (0.003–0.03)
Solar irradiance	0.12 (0.06–0.3)
Total	1.6 (0.6–2.4)

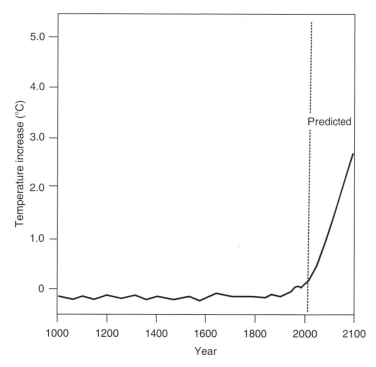

Fig. 2.5. Predicted temperature increases if carbon dioxide emissions are not restricted. (Redrawn from IPCC, 2006.)

and drought. Thus, it can be concluded that 'the balance of evidence suggests that there is discernable human influence on global climate' (IPCC, 2007).

The melting of the sea ice and glaciers will increase sea level by 0.5 m which will directly affect people living in low-lying areas such as the delta regions of Egypt, China and Bangladesh where 6 million people live below the 1 m contour.

The evidence of global warming has come from a very wide range of studies rather than the monitoring of temperature, and some of the trends which have indicated global warming are as follows (IPCC, 2007):

- Global mean temperature in 1900–2000 increased by 0.6°C (Fig. 2.5).
- Lowest 8 km of atmosphere in 1979–2001 increased by 0.05°C.
- Ice and snow cover decreased by 10% in 1960–2001.
- Mean sea level has risen 0.1–0.2 m in 1900–2000.
- More frequent changes in the El Niño currents in 1970–2000.
- Increases in the number of cyclones.
- Decrease in glaciers by 10%.
- Decrease in sea ice in 1973–1994.
- Increase in growing season in 1981–1991 by 12 days.
- Increasing heavy rain in 1900–1994.
- Increase of 5–20% in Antarctic snowfall in 1980–2000.

The possible social and economic consequences of global warming as given by the Stern Report are shown in Fig. 2.6. It is clear that even if some of the predictions prove true, the production of greenhouse gases needs to be reduced as soon as possible.

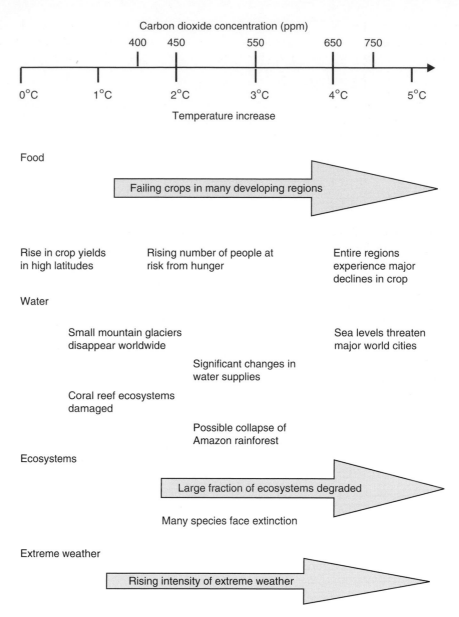

Fig. 2.6. Probable consequences of global warming in relation to carbon dioxide and temperature levels. (Redrawn from the Stern Report, Stern, 2006.)

Sources of Greenhouse Gases

Nitrous oxide (N₂O)

Nitrous oxide is an active greenhouse gas found at a very low concentration of 310 ppbv (parts per billion by volume) in the atmosphere. On a molecule-to-molecule basis nitrous oxide is 200 times more effective than carbon dioxide in absorbing infrared radiation and is also involved in the degradation of ozone. The gas is produced naturally by the denitrification of nitrate by microbial activity in soil and sea. Nitrous oxide is produced in a sequence of reactions leading from nitrate to nitrogen gas and is shown below:

$$NO_3 > NO_2 > NO > N_2O > N_2 \tag{2.1}$$

The sources of nitrous oxide are given in Table 2.4. The addition of nitrogen-based fertilizer to soils increases the rate of denitrification. Nitrous oxide is mainly lost in the stratosphere by photodegradation:

$$2N_2O + h\upsilon(\text{light}) = 2N_2 + O_2 \tag{2.2}$$

Methane

Methane (CH_4) is released from natural sources such as wetlands, termites, ruminants, oceans and hydrates (Table 2.5). Although the concentration of methane is

Table 2.4. Sources of nitrous oxide N_2O. (Adapted from Houghton *et al.*, 1990.)

Source	Nitrous oxide Mt/year	%
Natural		
Oceans	1.4–3.0	17.2
Tropical soils	2.5–5.7	32.8
Temperate soils	0.5–2.0	11.5
Anthropogenic		
Biomass burning	0.2–1.0	5.8
Cultivation	0.03–3.5	20.1
Industry	1.3–1.8	10.3
Cattle and feed	0.2–0.4	2.3

Table 2.5. Sources and sinks for methane. (Adapted from Houghton *et al.*, 1990.)

Source	Methane Mt/year
Natural	
Wetlands	115
Rice paddies	110
Ruminants	80
Biomass burning	40
Termites	40
Oceans	10
Freshwaters	5
Anthropogenic	
Gas drilling, venting	45
Coal mining	40
Hydrate distillation	5
Total	490
Removal	
Soil	30
Reaction with hydroxyl in atmosphere	~500

over 200 times lower than carbon dioxide, it is 21 times more effective at adsorbing infrared radiation than carbon dioxide. Methane levels are over twice what they were in pre-industrial times and have been increased by human activities such as rice cultivation, coal mining, waste disposal, biomass burning, landfills and cattle farms (Fig. 2.7). Ruminants can produce up to 40 l of methane per day. Methane is mainly removed from the atmosphere through reaction with hydroxyl radicals where it is a significant source of stratospheric water vapour. The remainder is removed through reactions with the soil and loss into the stratosphere.

Methane hydrates have been proposed as a potential source of energy but the exploitation of these deposits has its problems. It has been estimated that methane hydrates represent 21×10^{15} m³ of methane (11,000 Gt carbon) (Kvenvolden, 1999). The carbon content of these hydrates is greater than that contained in all the fossil fuels (Fig. 2.8) (Lee and Holder, 2001).

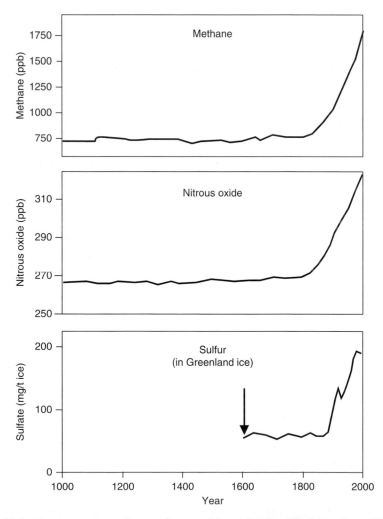

Fig. 2.7. Global increases in methane, nitrous oxide and sulfur. (Redrawn from IPCC, 1996.)

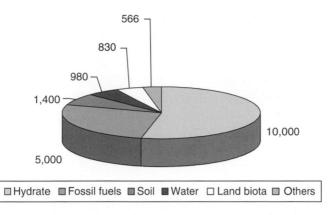

Fig. 2.8. The carbon content (Gt) of methane hydrates compared with other sources. (Redrawn from Lee and Holder, 2001.)

As a result, any controlled release of methane from the hydrate deposits may have a significant effect on global warming. There is increasing evidence that major releases of methane from hydrates have occurred in the past and have been associated with warming events, although insufficient methane may have been released to be responsible for the full rise in temperature (Glasby, 2003). One consequence of global warming may be the dissociation of some of the shallow hydrate deposit, further increasing global warming. Slow release of methane in the sea would result in its oxidation before reaching the surface but large-scale sediment slumping, such as the Storegga slide off Norway displacing 3900 km³ of sediment, may release huge quantities of methane.

Carbon dioxide

The carbon dioxide concentration in the atmosphere is low (368 ppmv; 0.03%) compared with oxygen and nitrogen but it is a greenhouse gas and is responsible for 55% contribution to global warming. There is a continual flow between the atmosphere and organic and inorganic carbon in the soils and oceans (Fig. 2.3). Plants on land and in sea fix carbon dioxide in photosynthesis and this is balanced by carbon dioxide produced by respiration of animal and plants and microbial decomposition of biological materials. Carbon dioxide is also locked up in plant and animal debris in soils and the oceans act as a very large sink where carbonate rocks and reefs also store carbon.

Over many millennia some of the plant and animal debris have been converted by high pressure and temperature into fossil fuels, oil, gas and coal. It is the burning of fossil fuels that is altering the balance of the atmospheric carbon dioxide.

Annual carbon dioxide emissions from the use of coal, gas and oil were above 23 Gt in 2000 having risen from 15.7 Gt in 1973 and 0 in pre-industrial times (IEA, 2002). Carbon dioxide emissions depend on energy and carbon content of the fuel, which ranges from 13.6 to 14.0 Mt C/EJ for natural gas, 19.0 to 20.3 for oil and 23.0 to 24.5 for coal (Wuebbles *et al.*, 1999).

The human activities that are responsible for greenhouse gas emissions are given in Fig. 2.9, from which it is clear that the energy sector dominates production.

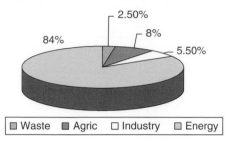

| □ Waste | ■ Agric | □ Industry | ■ Energy |

Fig. 2.9. Anthropogenic sources of greenhouse gases. (From Quadrelli and Peterson, 2007.)

Agriculture produces in the main the greenhouse gases methane (CH_4) and nitrous oxide (N_2O) from cultivation and livestock. When considering carbon dioxide, we see that the energy sector produces 95% carbon dioxide when Annex 1 countries are surveyed, with 4% methane and 1% nitrous oxide. The carbon dioxide emissions of sectors of the energy sector are given in Fig. 2.10. This use of fossil fuels appears to be responsible for the rapid increase in atmospheric carbon dioxide since the 1800s, and the IPCC predict that the carbon dioxide levels will continue to increase to values of 700 ppm by the year 2100 if nothing is done (Fig. 2.11). A number of scenarios have been developed by the IPCC based on various assumptions on the degree of reduction in greenhouse emissions. The carbon dioxide emitted from the electricity sector is the largest followed by that from transport. Although coal only represents 25% of the fossil fuels used for electricity generation, it generates more carbon dioxide as it contains a higher carbon content (Fig. 2.12). Figure 2.12 shows the fuels used in the global energy supply and the proportion of carbon dioxide that these produce. It is clear that coal use produces the greater proportion of carbon dioxide.

The carbon dioxide emissions in the 1990s were estimated to be 6.3 ± 0.4 PgC/year (6.3 Gt), which resulted in an increase in atmospheric carbon dioxide of 3.2 ± 0.1 Pg/year while the remainder was adsorbed by the oceans and land (Glasby, 2006). This equates to 25.2 Gt of carbon dioxide per year. There have been a number of estimates and calculations on the levels of carbon dioxide that would be obtained if various reduction scenarios were implemented (Fig. 2.11) (IPCC, 2007). The present-day global reserves of oil, gas and coal (Tables 1.9 and 1.10) are about 1091–1268 Gt carbon. If these reserves were all used, the final atmospheric carbon dioxide level would be 2000–2200 ppm (IPCC, 2006; Glasby, 2006). One of the targets is to hold atmospheric

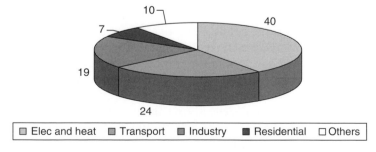

| □ Elec and heat | ■ Transport | ■ Industry | ■ Residential | □ Others |

Fig. 2.10. A total of 26.5 Gt carbon dioxide emissions in the world by sectors. (From Quadrelli and Peterson, 2007.)

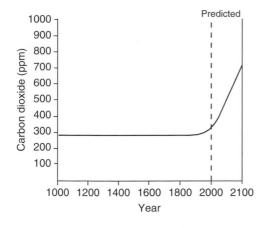

Fig. 2.11. Present and predicted increase in global carbon dioxide. (Redrawn from IPCC, 1996.)

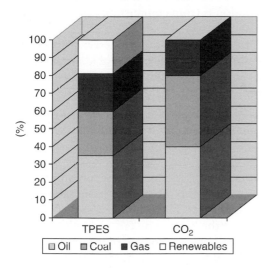

Fig. 2.12. The world's sources of energy and their carbon dioxide emissions. TPES: total primary energy supply. (From Quadrelli and Peterson, 2007.)

carbon dioxide at 450 ppm, which represents an increase of 70 ppm from present values equivalent to 13.2 PgC or 3.8% of fossil fuel reserves. If the target is 750 ppm, an increase of 350 ppm, this would be equal to 66 PgC (19%) (Glasby, 2006).

Other Fossil Fuel Pollutants

The burning of coal and oil also produces a number of harmful compounds other than carbon dioxide including carbon monoxide, sulfur dioxide and oxides of nitrogen.

Nitrogen oxides (NO$_x$)

NO$_x$ is the term given to the two oxides of nitrogen, nitric oxide (NO) and nitrogen oxide (NO$_2$) where nitric oxide predominates. These contribute towards acid rain as they are converted into nitric acid in the atmosphere. The natural sources of nitric oxide (NO) and nitrogen oxide (NO_2) are soils, ammonia oxidation and lightning (Table 2.6).

High temperatures such as those generated by lightning can produce nitrogen oxides. At high temperatures nitrogen reacts with oxygen by a number of mechanisms, including the Zeldovitch mechanism:

$$O_2 + N_2 \leftrightarrow NO + N \tag{2.3}$$

$$N + O_2 \leftrightarrow NO + O \tag{2.4}$$

$$N + OH \leftrightarrow NO + H \tag{2.5}$$

$$N_2 + O + M \leftrightarrow N_2O + M \text{ (M is an ion)} \tag{2.6}$$

The reaction yields a mixture of NO and N_2O, with NO dominating at ~90%. The high temperature reaction occurs in vehicle engines, aircraft engines and during biomass burning. It is clear that a major contribution is from combustion of fuels including hydrogen in engines of various types.

Most oxides of nitrogen are oxidized to NO_2 in the atmosphere which in the presence of water produces nitric acid. The nitric acid contributes to acid rain and the deposition of nitrogen into rivers and lakes can cause eutrophication. In humans NO_x has been implicated in an increased susceptibility to asthma.

Sulfur dioxide

The concentration of sulfur dioxide is less than 1 ppb in clean air to 2 ppm in highly polluted areas, with levels typical at 0.1–0.5 ppm. Sulfur dioxide is a respiratory irritant, which can affect human health and damage plants. There are a number of natural and anthropogenic sources of sulfur dioxide, but the latter is by far the largest source. Marine phytoplankton produces dimethyl sulfide, which is converted into sulfur dioxide in the atmosphere, hydrogen sulfide is formed by anaerobic decay and volcanoes emit sulfur dioxide. Most of the sulfur dioxide produced by human activity is from the burning of fossil fuels and the sources in fuels are listed on the following page:

Table 2.6. Sources of nitrogen oxide.

Source	Nitrogen oxides (NO_x) Mt/year
Natural	
Soil	18.1
Ammonia oxidation	10.2
Lightning	16.4
Anthropogenic	
Fuel combustion	65.1
Aircraft	3.0
Industry	4.0
Biomass burning	36.8

- Oil products contain between 0.1 and 3% sulfides.
- Natural gas can contain hydrogen sulfide which is often removed before use.
- Coal contains between 0.1 and –4% sulfur as inorganic iron pyrites and organic thiophenes.

Burning fossil fuels in power stations has been the main source of sulfur dioxide, as can be seen from the typical emissions from a power station (Table 2.7). The main emissions include carbon dioxide, sulfur dioxide from the sulfur compounds in the coal, and nitrous oxides from the nitrogenous compounds. In the atmosphere sulfur dioxide is rapidly oxidized to sulfuric acid. The pH of clean rainwater is about 5.6 due to dissolved carbon dioxide. However, rainwater in the presence of pollutants sulfur dioxide and nitrous oxides forms sulfuric, sulfurous and nitric acid which can reduce the pH to 1. These acids have a short residence time in the atmosphere, returning to the surface as rain. The acid rain has an effect on water bodies, vegetation and buildings. The problem of changes in water and soil pH has been of concern since the late 1960s.

Acidification of waters causes an increase in the leaching of toxic metal ions into the water and also changes the flora and fauna. Acid rain has been blamed for the death of trees in a number of forests in Scandinavia and the USA. The effect is perhaps indirect as the change in pH of the soil may leach out toxic metals and change the uptake of ions by plants.

Acid rain has an effect on buildings, particularly those made from limestone. Concern about acid rain was sufficient to initiate legislation to reduce emissions.

At present, methods of reducing sulfur dioxide emissions are as follows, and examples are given in Table 2.8:

- Reduction in fuel use by improvements in combustion and reduction in energy loss. Combined cycle gas turbine is one such system where the hot gases from the turbine are used to generate steam, which is then used to run a conventional turbine. This gives an efficiency of 53% compared with 35% for the conventional

Table 2.7. Typical emissions from a coal-fired power station prior to flue gas treatment. (Adapted from Roberts *et al.*, 1990.)

Chemical	Concentration
Air (oxygen depleted)	~80%
Water (H_2O)	~4.5%
Carbon dioxide (CO_2)	~12%
Carbon monoxide (CO)	40 ppm
Sulfur dioxide (SO_2)	1000–1700 ppm
Sulfur trioxide (SO_3)	1–5 ppm
Nitric oxide (NO)	400–600 ppm
Nitrogen dioxide (NO_2)	~20 ppm
Nitrous oxide (N_2O)	~40 ppm
Hydrochloric acid (HCl)	250 ppm
Hydrofluoric acid (HF)	<20 ppm
Particulates	<115 mg/m^3
Mercury (Hg)	3 ppb

Table 2.8. Annual emissions (in tonnes) from typical conventional UK 2000 MW power stations[d] compared with more efficient alternatives. (Adapted from IEE, 2002.)

Emissions	Coal-fired conventional	Oil-fired conventional	Gas-fired combined-cycle gas turbine
Particulates	7,000	3,000	Nil
Sulfur dioxide	150,000 (15,000[a] & 75,000[b])	170,000	Nil
Nitrogen oxides	45,000 (30,000[c])	32,000	10,000
Carbon monoxide	2,500	3,600	270
Hydrocarbons	750	260	180
Carbon dioxide	11,000,000	9,000,000	6,000,000
Hydrochloric acid	5,000	Nil	Nil
Ash	840,000	Nil	Nil

[a]Flue gas desulfuration.
[b]Low sulfur coal.
[c]Low NO_x burners.
[d]A power station of this size will produce 12 TWh/year.

coal-fired stations with sulfur dioxide removal. Clean coal systems have also been introduced using fluidized bed combustion and gasification.
- The use of low sulfur fuels. Inorganic sulfur compounds in coal, such as pyrites, can be removed by catalytic hydrodesulfuration. Organic sulfur compounds, principally thiophenes, can be removed by microbial action (McEldowney *et al.*, 1993). Replacing coal with natural gas also reduces sulfur emissions as natural gas contains little sulfur.
- Sulfur compounds can be removed from the flue gas and the most common is the limestone/gypsum method, where the flue gas is treated with calcium carbonate slurry (limestone). The calcium carbonate reacts with sulfur dioxide to yield insoluble calcium sulfate (gypsum) which precipitates and can be removed.

Photochemical smog

The burning of fossil fuels also produces particulates, soot and black smoke from vehicles and power stations. Power stations now control these types of emissions and there are regulations on the particulate emissions from vehicles. Before the Clean Air Act in the UK coal fires were responsible for the creation of smog, a mixture of fog and smoke, in large cities but these do not occur now. However, photochemical smogs do occur at the present time where NO_x, mainly NO, and unburnt hydrocarbons build up due to high traffic density in cities such as Mexico City, Bangkok and Los Angeles. Nitric oxide (NO) is converted into nitrogen dioxide, and nitrogen dioxide, which catalyses photochemically the production of ozone, produces a corrosive smog.

Conclusions

It is clear from reports such as the IPCC (2007) and Wuebbles *et al.* (1999) that global warming is occurring and mankind's activity is responsible for this. The causes of the warming are the greenhouse gases, largely carbon dioxide, which stop long-wave radiation from leaving the earth. These gases are responsible for maintaining the world's temperature at an average of 14°C, whereas without these gases the temperature would be at −19°C. However, mankind's burning of fossil fuels releases greenhouse gases, which causes an increase in their levels in the atmosphere and which in turn increases the global temperature. Some still argue that global warming is nothing more than natural variation, but data gathered from a large number of scientific fields indicate that a change in temperature is occurring and the change is a consequence of mankind's activity. The link between changes observed and the Industrial Revolution is difficult to refute. The consensus now is that the climate is being affected and the consequences may be severe.

3 Mitigation of Global Warming

Introduction

Greenhouse gases in the atmosphere are increasing, in particular carbon dioxide, from the steady values found before 1850. The increase in greenhouse gases appears to be due to the burning of fossil fuels, which has fuelled industrialization. In addition, the demands for energy are increasing as more countries become industrialized. If this increase in greenhouse gases continues unchecked, the consensus is that the world's climate may be adversely affected (Stern, 2006; IPCC, 2007). The major sources of global energy are the fossil fuels, the supply of which is finite. Therefore, given these factors, considerable efforts have been made to develop non-fossil energy sources. The sources should be both sustainable and renewable, and reduce or eliminate greenhouse gas emissions to the atmosphere.

Renewable energy means an energy source that can be continually replaced, such as solar energy and plant materials, where the energy is obtained from the sun or stored as a consequence of photosynthesis. However, there are some restrictions to renewable energy sources as these should not be depleted faster than the source can renew itself.

Sustainable development focuses on the long-term development to allow a switch from the use of finite resources to those which can be renewed. Sustainability has also become a political movement involving groups working to save the environment. Another term used for non-fossil energy sources is 'carbon-neutral' which means that either the energy production yields no carbon dioxide, such as solar and nuclear power, or the process only releases carbon dioxide previously fixed through photosynthesis. In determining the carbon dioxide reduction for renewable energy sources, life-cycle analysis will determine the fossil fuel input into the production of the fuel and carbon dioxide produced. These must be taken into consideration when the carbon dioxide savings are determined for some biofuels, which may be less than 100% carbon-neutral.

Urgent action is needed if the atmospheric greenhouse gases are to be stabilized at levels that would avoid damaging climate changes. This chapter covers the possible options available for the stabilization of greenhouse gases. In 2005, the world's emissions of greenhouse gases were 33.7 Gt which included 26.5 Gt of carbon dioxide (Quadrelli and Peterson, 2007). Table 3.1 lists the top 25 carbon dioxide-producing countries.

The USA is the largest producer of carbon dioxide, but China is predicted to overtake the USA by 2025 (World Resources Institute, 2006). Neither country has signed the Kyoto Protocol. Both India and China have had rapid economic growth in the past few years and will account for 45% of new energy demand by 2030 (IEA, 2007). Coal is being used to generate electricity in both China and India which will increase greenhouse gas emissions more rapidly than other fossil fuels.

Table 3.1. Top greenhouse-gas-emitting countries. (From World Resources Institute, 2006; UNEP, 2006 and UNFCCC, 2008.)

Country	GHG[b] emissions Mt CO_2 equiv.	% of world	CO_2 emissions[b] Mt CO_2 equiv.	% of world
USA	6,928	20.6	5,665	24.2
China	4,938	14.7	2,997	12.8
Russia	1,915	5.7	1,506	6.4
India	1,884	5.6	937	4.0
Japan[a]	1,317	3.9	1,155	4.9
Germany[a]	1,009	3.0	833	3.6
Brazil	851	2.5	303	1.3
Canada[a]	680	2.0	527	2.2
UK[a]	654	1.9	531	2.3
Italy[a]	531	1.6	426	1.8
South Korea	521	1.5	434	1.9
France[a]	513	1.5	373	1.6
Mexico	512	1.5	360	1.5
Indonesia	503	1.5	269	1.1
Australia[a]	491	1.5	329	1.4
Ukraine[a]	482	1.4	301	1.3
Iran	480	1.4	292	1.2
South Africa	417	1.2	296	1.3
Spain[a]	381	1.1	385	
Poland[a]	381	1.1	386	–
Turkey	355	1.1		–
Saudi Arabia	341	1.0	261	1.1
Argentina	289	0.9		–
Pakistan	285	0.8	–	–
Top 25	27,915	83		
World	33,666	100		

[a]Countries included in Annex 1, which also includes some countries with economies in transition.
[b]GHG includes CO_2, CH_4, N_2O, CFCs for the year 2000; CO_2 data for 2000.

Kyoto Protocol

The concerns on global warming started with the UN's world's first climate conference in Geneva in 1979, where it was agreed to set up a panel to review the data on global warming. The panel set up was named the Intergovernmental Panel on Climate Change (IPCC), which was initiated in 1988 and its first report was published in 1990. The timeline of the UN conferences and the Kyoto Protocol is shown in Fig. 3.1. Throughout the period that the IPCC has gathered data, there have been sceptics who have dismissed global warming as natural variation. For this reason the IPCC reports have been cautious in coming to any conclusion.

In the first IPCC report, it was concluded that the planet was warming and that human activity was possibly responsible. In the second IPCC report, in 1995, it concluded that the balance of evidence indicated global warming was occurring. At the second United Nations Conference on Environment and Development (UNCED)

START IN 1979 WITH WORLD FIRST CLIMATE
CONFERENCE IN GENEVA
|
IPCC STARTED IN 1988
|
1988 TORONTO SCIENTIFIC CONFERENCE
|
1990 FIRST IPCC REPORT
Planet is warming and human
activity responsible
|
1992 SECOND 'EARTH SUMMIT' UNFCCC CONFERENCE IN
RIO DE JANEIRO
|
1995 SECOND IPCC REPORT
Balance of evidence suggest global warming
1995 COP-1 MEETING IN BERLIN
|
1997 COP-3 MEETING IN KYOTO WHERE THE 'KYOTO PROTOCOL'
FORMULATED
5% reduction in 1990 values by 2010
|
1998 UNFCCC COP4 BUENOS AIRES
|
2001 THIRD IPCC REPORT
|
2002 THIRD 'EARTH SUMMIT' JOHANNESBURG
|
2004 UNFCCC COP-10 BUENOS AIRES
Russia ratifies Kyoto
|
2005 KYOTO IN FORCE

Fig. 3.1. Kyoto Protocol timeline.

'Earth Summit' in Rio de Janeiro in 1992, the convention, named the UN Framework Convention on Climate Change (UNFCCC), was adopted which came into force in 1994. At the Rio meeting, it was concluded that carbon dioxide emissions needed to be reduced, and a Conference of the Parties (COP) was set up to oversee this reduction. A series of meetings of the COP to the UNFCCC have discussed the problems of global warming. Discussions by the developed countries started at COP 1 in Berlin, and after 2.5 years of negotiations the Kyoto Protocol was agreed at the COP 3 conference in Kyoto in 1997. Although 84 countries initially signed the Protocol many were reluctant to sign before the details were clear. To be put into effect the Protocol had to be ratified by the developed countries, which produce 55% of the global carbon dioxide. To date some 182 countries have ratified the Protocol, and with Russia's recent ratification the Kyoto Protocol came into operation in February 2005. Ratification means that countries in Annex 1, which are developed or with an economy in transition to a developed state, have agreed to cap their emissions of greenhouse gases. The Protocol commits these countries to individual, legally binding targets to limit or reduce greenhouse gas emissions. For example, carbon dioxide emissions were to be reduced by an average of 5.2% below 1990 levels by 2008–2012 and a reduction of 20% by 2010. For some of the countries reductions will need to be quite large as emissions have continued to rise between 1990 and the present.

Most of the members of the Organisation for Economic Cooperation and Development (OECD) plus the states of Central and Eastern Europe are included in Annex 1 (Fig. 3.2). Under the Kyoto Protocol, the EU 15 was to be regarded as a single unit with a target of 8% reduction, but each country was also assigned a separate value. For example, although the EU figure was 8%, Germany was set a target of 21% reduction and the UK 12.5%, whereas others had an increase such as Spain at +15%. Countries joining the EU later in 2004–2007 have been assigned their own individual Kyoto targets.

The Kyoto Protocol is essentially a 'cap and trade' system, which includes a flexible mechanism that allows the purchasing of greenhouse reductions from other countries. Trading is allowed under the control of the Clean Development Mechanism (CDM), which means that any greenhouse gas reduction scheme in non-Annex 1 countries will earn Carbon Credits which can be sold to Annex 1 countries. The trading scheme was put into place because there were fears that the costs of emission reductions would be too expensive for Annex 1 countries and as an incentive for non-annex countries to reduce emissions. The EU created the EU Emissions Trading Scheme (EU ETS) and the UK its own voluntary scheme (UK ETS). Figure 3.2 gives the reduction in emissions agreed under the Kyoto Protocol by Annex 1 countries and the percentage change achieved by 2005. The changes in carbon dioxide emissions from three of the large producers not in the Kyoto Protocol are +15.8% for the USA, +47% for China and +55% for India. Recently India has ratified the Kyoto Protocol, and is now a member of the Annex 1 countries. The reason for the large increases in emissions from China and India is their rapid industrialization since the late 1990s.

The targets for the Kyoto Protocol are to be calculated as an average over a 5-year period. Progress in reduction in the three greenhouse gases – carbon dioxide, methane and nitrous oxide – will be measured against the values for 1990. The chlorofluorocarbons (CFCs), another greenhouse gas, have been dealt with under the 1987 Montreal Protocol, where their use was banned.

Reduction in the Global Greenhouse Gases

The objective of the UNFCCC is to stabilize greenhouse gases at an atmospheric level that would avoid damaging the environment completely and reduce further global warming. The Annex 1 countries are required to adopt climate change policies and measures to reduce emissions of greenhouse gases. The IPCC has developed a number of scenarios in order to predict carbon dioxide emissions by 2030. Some of these are shown in Fig. 3.3.

In the high growth scenario with no reduction in greenhouse gas emissions, carbon dioxide will reach 42 Gt/year by 2030, mainly from the economies of the USA, China, Russia and India reaching around 700 ppm carbon dioxide in the atmosphere. In the alternative scenario, where greenhouse gases are reduced, carbon dioxide emissions peak at 2012 and begin to decline after 2015. This would represent a final atmospheric carbon dioxide concentration of 550 ppm which corresponds to a temperature rise of 3°C.

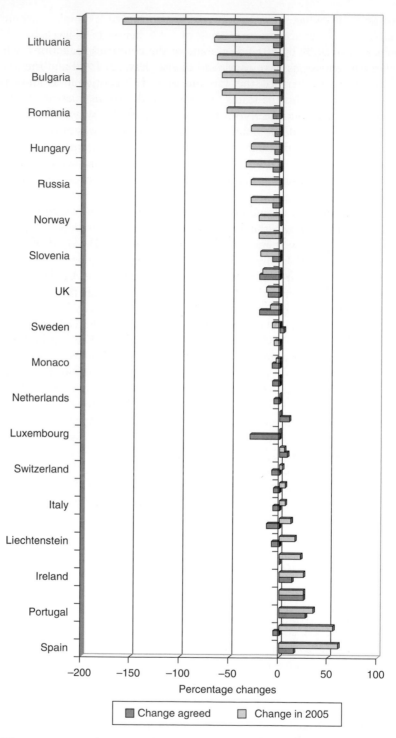

Fig. 3.2. The percentage changes achieved by Annex 1 countries by 2005 and those agreed in the Kyoto Protocol for 2008–2012. (From UNFCCC, 2008.)

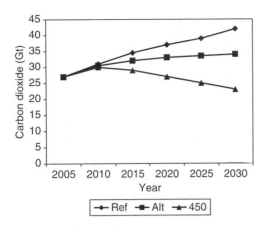

Chapter 3

Fig. 3.3. Three scenarios for the stabilization of atmospheric carbon dioxide. The high growth (♦); alternative scenario (■); 450 stabilization (▲). (Redrawn from IEA, 2007.)

This may be achieved by structural changes in the economy and changes to the fuels used in the largest emitting countries. The 450 ppm stabilization is the lowest in the IPCC scenarios, and requires a drop in carbon dioxide emissions to 23 Gt by 2030, which corresponds to a temperature change of 2.4°C. This would require considerable changes in both Annex 1 and non-annex countries, many of which would be costly.

As can be seen in Fig. 3.4, although many of the Annex 1 countries have reduced their greenhouse gas emissions, the global emissions have continued to rise since the Kyoto Protocol came into being.

It is clear that those developing nations with high emission levels will need to implement emission control if the atmospheric concentrations of greenhouse gases are to be stabilized.

Reduction in EU Greenhouse Gas Emissions

In 2001 a number of directives were produced to encourage renewable energy sources. Directive 2001/77/EC set a target for each member state for the proportion of electricity produced from sustainable resources and later the ten new members also

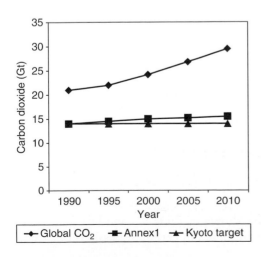

Fig. 3.4. Global, Annex 1 and Kyoto target carbon dioxide emissions now and predicted in Gt. (From IEA, 2005a.)

Table 3.2. Share as a percentage of possible alternative fuels for transport in the EU by 2020. (From Demirbas, 2008.)

Year	Biofuels	Natural gas	Hydrogen	Total
2010	6	2	–	8
2015	7	5	2	14
2020	8	10	5	23

set up national targets. Directive 2003/30/EC covered the promotion of biofuels for transport, and member states are permitted to reduce excise duties on biofuels (Directive 2003/96/EC).

Under what is considered an optimistic view, Table 3.2 shows the possible share of alternative fuels in the EU by 2020 (Demirbas, 2008). In this case, natural gas use increases until 2030–2040, after which it declines. Biofuels show a steady increase up to 2050, and over this time period hydrogen overtakes biofuels and replaces natural gas.

Reduction in UK Greenhouse Gas Emissions

In 1990, the UK government published its first white paper on sustainability, followed by, in 1999, a paper, *A Better Quality of Life*. More recently the Department of Trade & Industry (DTI) produced a white paper, *Our Energy Future – Creating a Low Carbon Economy* in 2003 in which there was an aim to reduce greenhouse emissions by 60% by 2050 compared with the 1990 value.

As part of the measures to reduce carbon dioxide emissions, the UK has introduced a major policy, the Renewable Transport Fuel Obligation (RTFO), formerly Non-Fossil Fuels Obligation, in response to the EU Biofuels Directive (2003/30/EC). The government has enacted the Renewable Obligation Order, under which 10% of electricity generation should come from renewables by 2010. Most of the renewable-produced electricity is expected to come from wind and co-firing with biomass. In the UK, 6.7% of the electricity should be generated from biomass by 2006/2007 and 15.4% by 2015/2016. The current schemes under the Renewable Obligation Order include the following:

- Any biomass, including imported coconut and olive waste, can be co-fired until March 2009.
- Of the total produced, 25% of co-fired biomass must be from energy crops from April 2009 until March 2010.
- Also, 50% of co-fired biomass must be from energy crops from April 2010 until March 2011.
- In addition, 75% of co-fired biomass must be from energy crops from April 2011 to March 2016.
- Co-firing ceases to be eligible for renewable obligation certificates (ROCs) after March 2016.

The most recent figures given for the UK are shown in Table 3.3 and Fig. 3.5, where the greenhouse gases are quoted as millions of tonnes of carbon. Emissions of carbon

Table 3.3. Emissions of greenhouse gases in the UK since 1990 in million of tonnes (Mt) of carbon. (Adapted from Defra, 2007.)

	1990	1995	2000	2001	2002	2003	2004	2005	2006
Carbon dioxide	592.4	549.8	548.6	559.4	542.7	554.7	555.1	555.2	554.5
Methane	103.5	90.2	68.4	62.4	59.4	53.4	51.6	49.5	49.1
Nitrous oxide	63.8	53.0	43.6	41.5	40.1	39.8	40.6	39.8	38.3
CFCs and sulfur hexafluoride	13.8	17.2	11.7	11.5	11.7	11.8	10.3	10.6	10.4
Kyoto GHG basket[a]	770.8	709.0	671.4	674.4	653.8	659.5	657.9	655.5	652.3
Total allowing for EU ETS	770.8	709.0	671.4	674.4	653.8	659.5	657.9	628.4	619.0
Change from baseline 779.9 (%)	−1.2	−9.1	−13.9	−13.5	−16.2	−15.4	−15.6	−19.4	−20.6

[a]The Kyoto basket differs slightly from sum of gases because of change in LULUCF.

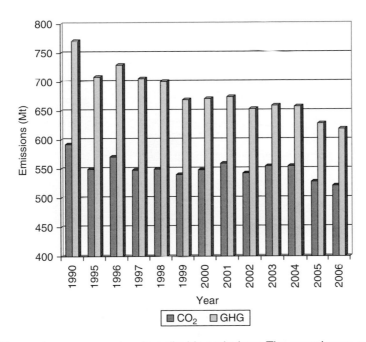

Fig. 3.5. UK greenhouse gas and carbon dioxide emissions. The greenhouse gases include carbon dioxide, methane, nitrous oxide, hydrofluorocarbons, perfluorocarbons and sulfur hexafluoride and allow for EU ETS in 2005 and 2006. (Redrawn from Defra, 2008.)

dioxide fell by 12% between 1990 and 2006, helped by emissions trading in 2005 and 2006 (EU ETS). The overall drop in greenhouse gases which include carbon dioxide, methane, nitrous oxide and the chlorofluorocarbons and sulfur hexafluoride was 20.6% (Table 3.3) in 2006, well below that required by the Kyoto Protocol. The factors contributing to this reduction are given as less coal and oil used and more gas and renewables used, particularly to generate electricity (Defra, 2008).

The RTFO also sets levels of biofuels, bioethanol and biodiesel to be added to overall fuels sales with a final value of 5% by volume. The levels of obligation are given in Fig. 3.6 in terms of volume. If the energy content of the fuel is used, the addition represents 5.75% because of the lower energy content of biofuels. However, the levels of biofuel use in the UK have suffered some difficulties and are still low (Table 3.4). The UK Report to the European Commission quoted these difficulties as 'the level of duty differential in the UK and the length of certainty the mechanism offers has proved insufficient to stimulate the level of investment in production capacity and infrastructure required to meet the Directive's objectives'.

The development of biofuels and the effects of policy can be seen in the contrasting positions in the UK and Germany. The markets for biofuels have been compared for the UK and Germany in terms of the directives, policies and standards (Bomb *et al.*, 2007). In 2002, the German government exempted all biofuels from excise tax until 2009 to encourage their adoption. Germany is now the largest producer of biodiesel in the EU, and after 2004 the excise duty was exempted for biodiesel blends. This has made biofuels competitive with fossil fuels in terms of cost. However, some groups in Germany have been critical of biofuels, including Friends of the Earth (Bomb *et al.*, 2007). At present the market for 100% biodiesel has shifted to trucks, and automobile manufacturers are making efforts to provide warranties for 100% biodiesel and blends for cars. In 2004 no ethanol was produced in Germany and any used was imported from Brazil, but a number of ethanol plants are now under

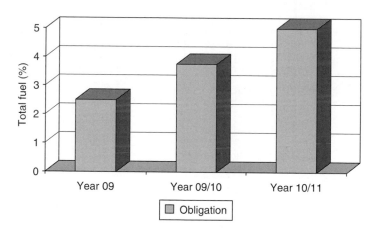

Fig. 3.6. The level of obligation on the addition of biofuels to all fuels set by the UK in 2006. (Redrawn from UK Report to European Commission Article 4, Department of Transport, 2006.)

Table 3.4. UK biofuel sales in the UK for 2005. (From UK Report to European Commission Article 4, UK Department of Transport, 2006.)

Fuel	Sales in 2005 (million litres)	% of total fuel sales	% total of total fuel energy content
Biodiesel	33	0.07%	0.06%
Bioethanol	85	0.17%	0.12%
Total	118	0.24%	0.18%

construction. Ethanol use in Germany has been encouraged with support of the farm sector, energy security and reduction in carbon dioxide emissions. Not only has there been an exemption of excise tax but additional factors such as set-aside premiums, partial state monopoly and price controls for agricultural products.

In contrast, in the UK, the excise tax on biodiesel was reduced in 2002 by £0.20 and ethanol by £0.20 in 2005. The excise tax still remained at £0.275 on both biofuels, which means that only biodiesel produced from used cooking oil is competitive in price with fossil fuels. However, production of biodiesel in the UK has increased and in 2008, 404,000t of biodiesel from a variety of oil sources and 55,000t of bioethanol from sugarbeet were produced.

The main driver for the introduction of biofuels in Germany has been the exemption of excise tax, which was not implemented in the UK, and explains the differences in production in both countries. The tax concession is expected to end in 2009 in Germany and it will be interesting to see how biofuels will compete after this time. The result of the policies or the lack of policies in the two countries explains the contrasting adoption of biofuels, and the conclusions of a study of the two countries are as follows (Bomb *et al.*, 2007):

- Consumers will buy fuel on price, rather than for green credentials.
- National government commitment for the establishment of a biofuel industry is required.
- Low-level blending is the easiest method of introducing biofuels, but sufficient volumes for anything more than 5% addition are not available.
- Niche markets are available for biofuels in areas of environmental sensitivity.
- Oil companies are more supportive of biodiesel than bioethanol.
- Environmental impacts and carbon balances vary between biofuels, and there have been some questions about bioethanol.
- A fuel certification system is needed for sustainability and fuel composition.
- Support for biodiesel and bioethanol does not stop other technologies from being developed.

Reduction in Greenhouse Gas Emissions

Different measures have been proposed to reduce global warming caused by the burning of fossil fuels. However, the various bodies involved differ in defining the steps that need to be taken. Seven steps have been suggested by Mathews (2007), which require no further technological advances to be implemented, and should reduce emissions by 70% by 2050.

These steps are as follows and involve both carbon taxes and permits:

1. A global carbon pricing regime based on carbon taxes and permits.
2. Global satellite monitoring of greenhouse gas emissions.
3. Compensating developing countries for preserving rainforests.
4. Creation of a global market for responsible biofuels.

Each region can certainly pursue biofuels adapted to conditions found there, such as rapeseed for biodiesel in Europe (Ryan *et al.*, 2006), but it is unrealistic to see temperate regions becoming self-sufficient in biofuels. It is far more expedient to open up the world market and to encourage trade in biofuels, both to accelerate the utilization of biofuels as a defence against global warming, and to encourage industrial development of tropical countries as the world's supplier of biofuels:

5. Creation and furtherance of markets for renewable electricity.
6. A global moratorium on building new coal-fired power stations.
7. Creation of global incentives for developing countries that are moving to adopt non-fossil-fuel industrial pathways.

Hoffert *et al.* (2002) stated: '[S]tabilizing climate is not easy. At the very least, it requires political will, targeted research and development and international cooperation. Most of all it requires the recognition that although regulation can play a role, the fossil fuel greenhouse effect is an energy problem that cannot be simply regulated away.'

All the above measures are a mechanism to implement the following measures which can be used to reduce global warming. These are simply:

1. Burn less fuel.
2. Sequester carbon dioxide produced.
3. Use renewable alternative fuels.

Burn Less Fossil Fuel

The reduction in fossil fuel use for electricity generation, heating/cooling and transport may involve a large number of measures, some of which are as follows:

- Increased engine efficiency.
- Increased power generation efficiency.
- Local electricity generation and distribution.
- Better home insulation.
- Fewer car and lorry journeys.
- Greater use of public transport.
- Greater use of biofuels.
- Alternative power systems.
- Changes in house design.
- Reduction in long-distance transport of material which can be sourced locally, and reduction in air miles.

Energy efficiency measures such as insulation, building design, light bulbs, stand-by default on televisions and other consumer electronics have been estimated to make a major contribution to reduction in energy use. The EU Emissions Trading Scheme and the Climate Change Levy should also encourage cost-effective energy saving, estimated at reducing carbon emissions by 6–9 Mt. Transport uses over 30% of the total energy, therefore continued increases in engine efficiency should give significant savings.

Distributed energy

Electricity is mainly generated in large power stations (2000 MW and above) and 75% of home heating comes from gas supplied through a nationwide network. While centralized systems deliver economies of scale, safety and reliability, the transfer of electricity to remote users loses 20.3 Mtoe, which is 8.7% of the total energy generated (Table 1.4). However, new and existing technologies, especially advances in gas turbines, have achieved maximum efficiency in small power plants of up to 10 MW (Poullikkas, 2005). This makes it possible to generate energy close to where it is used, which is known as 'distributed generation'. Distributed generation has been defined as 'a small scale power generation technology that provides electric power at a site closer to customers than central station generation and is usually interconnected to the transmission or distribution system' (Edinger and Kaul, 2000).

Distributed energy includes:

- All plants connected to a distribution network rather than transmission network.
- Small-scale plants that supply electricity to a building, industrial site or community.
- Microgeneration, small installations such as solar panels, wind turbines, biomass burners supplying one building or small community.
- Combined heat and power plants (CHPs), including large, community- or building sized and micro-CHP, replacing domestic boilers in homes.
- Non-gas sources of heat such as biomass, wood, thermal, solar or heat pumps for households and small communities.

These smaller systems can be more flexible and reduce the distribution losses incurred with a centralized system. At present less than 10% of electricity comes from microgeneration and CHP plants but these are increasing. The advantages of the distributed generation include:

- These plants can be more reliable.
- They are flexible in their energy source. These can handle renewable sources of power.
- They avoid transporting fuel long distances.
- Less power is lost in distribution.
- They enable the introduction of alternative power systems which can be intermittent such as wind power and photovoltaics.

These distributed systems could fundamentally change the way energy is supplied, and reduce transmission losses and fuel imports.

Alternative drive systems

Although not directly involving biofuels, the development of alternative drive systems that do not use fossil fuels for transport is important in the reduction in fossil fuel use and greenhouse gas emissions. There are a number of systems being tested including fuel cells, electric cars and hybrid systems.

Fuel cells

Fuel cells have had a long development, including use in the NASA Apollo programme in 1960, and since 1990 an experimental transportation system has been introduced. A fuel cell consists of two electrodes – the anode and cathode – divided by an electrolyte (Fig. 3.7). Hydrogen is run into the anode, where a platinum-coated

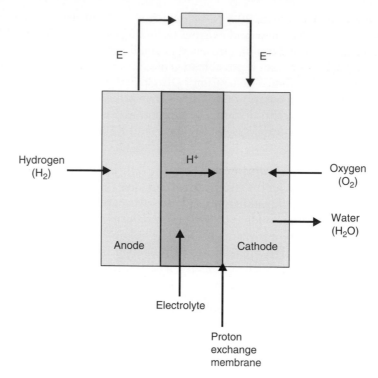

Fig. 3.7. Outline of a fuel cell.

proton-exchange membrane splits the hydrogen into hydrogen ions (protons) and electrons. The protons pass through the electrolyte to the cathode where they combine with oxygen, forming water. The electrons produce an external current which can be used to run an electric motor.

Fuel cells are classified by their operating temperature which is also determined by the electrolyte (Stambouli and Traversa, 2002). Table 3.5 gives some of the characteristics of fuel cells. Fuel cells can be combined in stacks, connected in series to produce the desired voltage. The number of fuel cells in a stack determines the voltage and the surface of each cell determines the current. Proton exchange and solid oxide fuel cells are the most advanced and have been fitted into experimental cars.

Two recent developments in fuel cell technology are the direct carbon fuel cell and the microbial fuel cell. In the direct carbon fuel cell, fine particles of carbon (10–1000 nm) are mixed with molten lithium, sodium, or potassium carbonate at 700–800°C (Cooper, 2006). The molten salt is introduced into the anode compartment and air to the cathode (Fig. 3.8). Electrons are carried from the carbonate to the cathode. Oxygen passes through a membrane which reacts with carbon, releasing electrons, forming carbon dioxide.

The microbial fuel cell derives energy from organic compounds metabolized by microorganisms. Figure 3.9 shows the layout of a microbial fuel cell. Microbes in the anode chamber oxidize substrates added to the chamber, generating electrons and protons as found in the chemical fuel cell. Carbon dioxide is formed but as organic substrates are used, the carbon dioxide released is only that fixed during photosynthesis.

Table 3.5. Characteristics of fuel cells.

Type	Electrolyte	Operating Temperature (°C)	Fuel
Proton-exchange membrane (PEMFC)	Polymer	50–200	Hydrogen
Phosphoric acid (PAFC)	Phosphoric acid	160–210	Hydrogen or hydrogen from methane
Molten carbonate (MCFC)	Molten salt, nitrate, sulfate carbonate	630–650	Hydrogen, carbon monoxide, natural gas, propane
Solid oxide (SOFC)	Zirconia	600–1000	Natural gas, propane, hydrogen
Solid polymer (SPFC)	Polystyrene	90	Hydrogen
Alkaline (AFC)	Potassium hydroxide, KOH	50–200	Hydrogen, hydrazine
Direct methanol (DMFC)	Polymer	60–100	Methanol

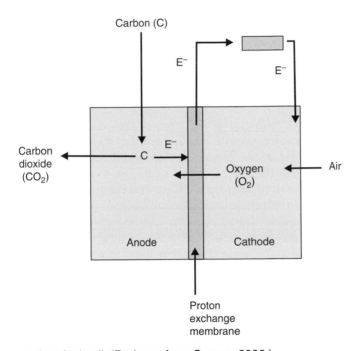

Fig. 3.8. Direct carbon fuel cell. (Redrawn from Cooper, 2006.)

The reactions are as follows when using acetate as a substrate:

$$CH_3COOH + 2H_2O \rightarrow 2CO_2 + 7H^+ + 8e^- \qquad (3.1)$$

At the cathode the protons react with oxygen:

$$O_2 + 4e^- + 4H^+ \rightarrow 2H_2O \qquad (3.2)$$

To extract electrons to the anode, mediators have to be added to the anode chamber. These mediators move across the microbial cell membrane where they are reduced

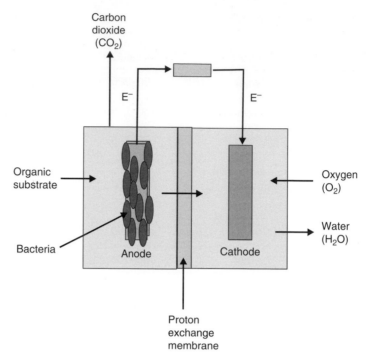

Fig. 3.9. Biological fuel cell.

and pass out of the cell, releasing the electrons to the anode. Mediators are dyes and metallorganics such as neutral red, methylene blue, thionine, Meldola's blue and 2-hydroxy-1,4-naphthoquinone. However, the instability of the mediators limits their use, but recently a group of bacteria, the anodophiles, have been isolated. These bacteria (including *Shewanella putrefaciens, Geobacteraceae sulfurreducens, Geobacter metallireducens* and *Rhodoferax ferrireducens*) attach themselves and transfer electrons directly to the anode (Du *et al.*, 2007). Some microbial fuel cells have been inoculated with bacteria mixtures such as sewage sludge and sediments which have the advantage of a wider substrate range. The amount of electricity provided by the microbial fuel cells is still very low, but they can be stacked and used to produce hydrogen, for wastewater treatment and as biosensors.

Alternative biological fuel cells have the microbial cells replaced with enzymes. This has the advantage of having a higher volumetric catalytic capacity, and it avoids toxic oxidation products. One of the fuels tested in an enzyme-based fuel cell is glycerol, one of the by-products of biodiesel production. Glycerol is a non-toxic, non-volatile, high-energy density substrate (6.3 kWh/l) for a cell containing the enzymes alcohol dehydrogenase and aldehyde dehydrogenase (Arechederra *et al.*, 2007).

Battery electric vehicles

The electric vehicle is ideal for use in cities as it emits no fumes and can be integrated into city-wide traffic systems where travelling distances are short. The electricity needed to charge the battery can be generated from renewable sources, which leads to a considerable reduction in carbon dioxide emissions. However, for long journeys battery technology and electricity storage is the key point. New battery systems such as Li-ion, nickel hydride and high temperature are under development (Van Mierlo *et al.*, 2006). Recently Mercedes has introduced a battery-powered Smart car. It is powered by a high temperature (260–330°C) sodium–nickel–chloride battery with an output of 15.5kWh. The battery can be recharged from the domestic supply and takes about 8h and gives the car a range of 50 miles. Considering the developments in battery technology driven by the mobile phone and computer industries it is likely that the range will be extended.

Hybrid vehicles

For long-range travel, hybrid vehicles appear to offer the best option. The hybrid is a combination of a battery-powered electric engine plus an internal combustion engine or fuel cell. The internal combustion engine can be used to charge the battery when in use. There are a number of options for hybrid drive trains and a number of cars available from a range of manufacturers which combine electric motors with a small petrol engine, notably the Toyota Prius.

Sequestration of Carbon Dioxide

One method of reducing or stabilizing atmospheric carbon dioxide levels is to trap and lock away carbon dioxide produced by the large carbon dioxide emitters such as power generation and cement works (Fig. 3.10). It is clear that electricity generation, particularly using coal, is the largest stationary source of carbon dioxide, followed by cement production and refineries.

Carbon sequestration can be defined as stable storage of carbon, but it has been suggested that storage in soil and plants cannot be regarded as stable as microbial degradation can lead to carbon dioxide release. However, the degree of permanence can vary greatly with carbon sequestered in soils with some components lasting up to 1000 years.

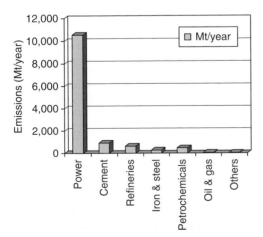

Fig. 3.10. Large global carbon dioxide producers.

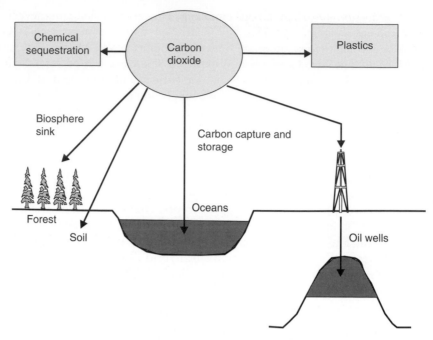

Fig. 3.11. Various methods that could be used for carbon dioxide sequestration from a large stationary source of carbon dioxide.

The possible methods of sequestering carbon dioxide from stationary sources are as follows (Fig. 3.11):

- Carbon capture and storage in the deep oceans, oil and gas reservoirs, aquifers and coal beds (geosphere sink).
- Planting more trees or reforestation (biosphere sink).
- Chemical sequestration (material sinks).
- Trapping the carbon dioxide in material such as plastics (material sinks).
- Agricultural practices (biosphere sink).

Carbon capture and storage

In order to sequester carbon dioxide it first has to be extracted from the flue gases of power stations, cement works and refineries in a process known as 'carbon capture and storage'. Carbon dioxide capture requires a relatively pure carbon dioxide stream for transport and storage. In some cases the carbon dioxide needs to be compressed, as mineral carbonization requires high pressures. There are three main methods of capturing carbon dioxide.

The first is post-combustion or process removal using absorption, adsorption, cryogenic and membrane technologies. Adsorption uses amine-based solvents or cold methanol. Carbon dioxide can also be removed by passing over activated charcoal or through special membranes (Feron and Jansen, 1995; Grimston *et al.*, 2001; Gronkvist *et al.*, 2006). Advanced methods are adsorption on to zeolites or activated carbon fibres.

The second is pre-combustion gasification of coal in a power station which produces a mixture of carbon monoxide and hydrogen. If this mixture is treated with steam, the carbon monoxide is converted to carbon dioxide and hydrogen. The hydrogen on combustion in the power station forms water and the carbon dioxide and water can easily be separated:

$$\text{Gasification} = CO + H_2 \tag{3.3}$$

$$\text{Water shift reaction } CO + H_2 + H_2O = H_2 + CO_2 \tag{3.4}$$

The third method is to use oxygen in the combustion process, which produces a flue gas of carbon dioxide and water. An alternative to oxygen addition is to use a metal oxide as an oxygen source, known as chemical looping.

Some examples of carbon dioxide sources suitable for carbon capture and storage are given in Fig. 3.11. Cement production is one example where carbon dioxide can be sequestered, and should be as cement production contributes 6–7% of the total global carbon dioxide released into the atmosphere. Cement factories are stationary and hence suitable for retrofitting of carbon dioxide sequestration. In cement production, calcium carbonate (limestone) is converted to lime (CaO) in a rotary kiln, releasing carbon dioxide. The lime is then heated at around 1450°C to form clinker, which when mixed with 5% gypsum forms cement (Gronkvist *et al.*, 2006). One tonne of cement releases about 500 kg carbon dioxide in its production. In addition, carbon dioxide is released from the fuel used to heat the kiln and has been estimated to be 275 kg carbon dioxide per tonne of lime, giving a total of 775 kg carbon dioxide produced per tonne of cement.

Whatever the process producing carbon dioxide, once the carbon dioxide has been separated it needs to be stored to keep it from reaching the atmosphere. The carbon dioxide can be stored using one of the following systems (Fig. 3.11):

1. Store underground in oil and gas reservoirs, deep saline aquifers, coal beds, active and uneconomical oil and gas reservoirs.
2. Hold in deep non-mineable coal formations and coal bed methane formations.
3. Store in deep oceans.

The storage underground could be part of enhanced oil recovery (EOR). At present 80% of recovered carbon dioxide is used in EOR and about 70 oil fields use this worldwide, sequestering some 31 million m³ of carbon dioxide per day. The retention times and capacities for carbon sequestration are given in Table 3.6.

The storage in the deep oceans has a number of possibilities, as the oceans contain 40,000 Gt carbon compared with 780 Gt in the atmosphere. Thus, the oceans are an immense carbon sink where captured carbon dioxide could be stored, and the options to put it in the oceans are as follows (Grimston *et al.*, 2001):

- Dry ice released into the sea from ships.
- Liquid carbon dioxide injected at depth of 1000 m from a ship or ocean-bottom manifold forming a rising droplet plume.
- A dense carbon dioxide–seawater mixture formed at a depth of 500–1000 m forming a sinking current.
- Liquid carbon dioxide introduced into a sea bed depression forming a stable lake at a depth of below 4000 m.

Table 3.6. Global capacity and residence time for the various carbon sinks. (Adapted from Grimston *et al.*, 2001.)

Sink	Capacity (GtC)	Retention time (years)
Oceans	1,000–10,000	Up to 1,000
Forestry	60–90	50
Agriculture	45–120	50–100
Enhanced oil recovery	20–65	10–50
Coal beds	80–260	>100,000
Oil and gas reservoirs	130–500	>100,000
Deep aquifers	30–650	>100,000

Forestry

All plants fix carbon dioxide during photosynthesis, which is released again when the plant dies and the plant material is degraded by microorganisms. The carbon dioxide fixed is used to synthesize storage compounds such as starch and oils, and cellular structural components such as cellulose and lignin. It is the structural components that are the slowest to degrade when the plant dies and these are the highest in woody plants.

It has been estimated that the amount of carbon taken up by vegetation was 3.2 Gt C/year, and 1.7 Gt C/year is lost mainly through deforestation, which gives a net increase of 1.5 Gt C/year. The carbon emissions from fossil fuels are 6.4 Gt C/year so that without any mitigation measures, some 23% of the carbon dioxide is removed. However, afforestation does not remove all the carbon dioxide produced and the atmospheric levels are still rising, but it does show the potential of biomass to sequester carbon dioxide. Table 3.7 indicates the potential of carbon dioxide sequestration by planting new forests, managing existing forests, managing crops, etc.

Woody plants have a carbon content of 0.54 kg carbon per kilogram of dry wood (Cook and Beyea, 2000). If tree growth is linear in the early years, the carbon dioxide removed can be calculated. The estimates for carbon dioxide sequestered by maize, switchgrass, short-rotation coppice willow and standing forest wood are shown in Table 3.8. Some of the carbon dioxide is only sequestered on a temporary basis as the switchgrass and short-rotation coppice will be burnt as a fuel, and other parts of the crops will be returned to the soil where they will be degraded. However, if the wood is used as construction material, the carbon dioxide will be locked up for considerably

Table 3.7. Potential for carbon dioxide sequestration by forests. (From Cannell, 2003.)

Region	Potential
Carbon storage capacity (Gt C)	
World	50–100
Europe	5–10
UK	0.3–0.5
Carbon sequestration rates 50–100 years (Mt C/year)	
World	1,000–2,000
Europe	50–100
UK	1–2

Table 3.8. Estimated carbon dioxide sequestration by various crops. (Adapted from Cook and Beyea, 2000.)

Biomass type	Carbon content (kg/kg; dry)	CO_2 sequestered (kg/kg)	Yield (t/ha)	CO_2 reduction (t/ha/year)
Maize	0.4	0.3	15–20	5.4
Switchgrass	0.4	0.4	15–20	7.4
Short-rotation coppice 3-year rotation	0.54	0.55	10–15	7.4
Forest wood 100-year rotation	0.54	0.14	–	–

longer. The carbon dioxide sequestered in forest material is less because growth is slower and forest regeneration, once harvested, is not always certain. The world is losing areas of dense forests either to building or agriculture, and reversal of this trend would help to reduce carbon dioxide levels in the short term.

Microalgal sequestration

The use of microalgae to sequester carbon dioxide was proposed in the past by a number of authors (Benemann, 1997; Sheehan *et al.*, 1998; Chisti, 2007; Skjanes *et al.*, 2007).

Microalgae have been proposed as systems for the sequestration of carbon dioxide (Sawayama *et al.*, 1995; Zeiler *et al.*, 1995; Ono and Cuello, 2006; Cheng *et al.*, 2006; de Morais and Costa, 2007a,b) and the production of biofuels (Chisti, 2007). The biofuels include biogas (methane) by anaerobic digestion of the biomass (Spolaore *et al.*, 2006), biodiesel from microalgal oils (Nagle and Lemke, 1990; Minowa *et al.*, 1995; Sawayama *et al.*, 1995; Miao and Wu, 2006; Xu *et al.*, 2006; Chisti, 2007), and biohydrogen (Fedorov *et al.*, 2005), and the direct use of algae in emulsion fuels (Scragg *et al.*, 2003).

Microalgae should be considered to have the following features:

- Higher photosynthetic efficiency than terrestrial plants.
- Rapid growth rate, doubling times of 8–24 h.
- High lipid content 20–70%.
- Direct capture of carbon dioxide, 100 t algae fix ~183 t carbon dioxide.
- Can be grown on a large scale.
- Will not compete with terrestrial plants in food production.
- Produce valuable products.
- Freshwater and marine species.
- Have a much better yield of oil per hectare: oil palm 5000 t/ha, algae 58,700 t/ha (Chisti, 2007).

Figure 3.12 shows a possible system for carbon dioxide sequestration and biofuel production using carbon dioxide from a stationary carbon dioxide source such as a power station. Microalgae can fix large quantities of carbon dioxide but it is likely that only a proportion of the carbon dioxide in the flue gases will be removed. Also the flue gases from power stations contain other gases which may affect the growth of microalgae. A number of studies have been carried out on the effect of flue gases on microalgae. *Nannochloris* sp. was shown to grow in the presence of 100 ppm

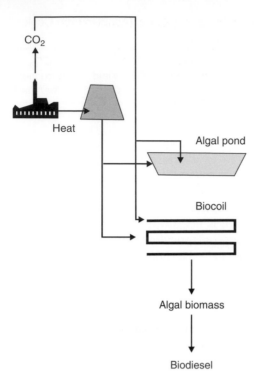

CO$_2$

Heat

Algal pond

Biocoil

Algal biomass

Biodiesel

Fig. 3.12. Possible sequestration of carbon dioxide from a power station and use of algal biomass to produce biodiesel.

nitric oxide (NO) (Yoshihara *et al.*, 1996). *Tetraselmis* sp. could grow in flue gas containing 185 ppm sulfur oxides, 125 ppm nitrogen oxides and 14.1% carbon dioxide (Matsumoto *et al.*, 1995). A *Chlorella* sp. was also found to grow in the presence of various combinations of sulfur and nitrogen oxides (Maeda *et al.*, 1995).

Chemical sequestration

Chemical sequestration involves the conversion of carbon dioxide to inorganic carbonates, and the use of carbon dioxide in the production of urea and plastics.

Agricultural practices

Agriculture uses a very large area of land and because of this size it is responsible for the emission of large quantities of greenhouse gases. Agriculture has been regarded as responsible for 25% of carbon dioxide, 50% of methane and 70% of the nitrous oxide released (Hutchinson *et al.*, 2007). The World Resources Institute (2006) gives the values of 27% carbon dioxide, 53% methane and 75% nitrous oxide. The greenhouse gas emissions from agricultural sources are given in Table 3.9. These emissions arise from the following:

- Fossil fuel use in cultivation, harvesting, etc.
- Nitrogen fertilizer use.

Table 3.9. Greenhouse gas emissions in Tg (10^{12} g) CO_2 equivalent from agricultural sources. (Adapted from Verge *et al.*, 2007.)

Gas	World 1990	World 2000
Nitrous oxide N_2O		
Agricultural soils	2240	2526
N fertilizers	374	444
Manure storage	181	190
Methane CH_4		
Enteric fermentation (cattle)	1836	1835
Manure management	200	206
Rice cultivation	845	898
Total GHG (Tg CO_2 equivalent)	5310	5656

- Rice production.
- Deforestation.
- Livestock (enteric fermentation).
- Manure.

Mitigation of greenhouse gases, sometimes known as stabilization, may require methods that are expensive, therefore low-cost options have been investigated. Agriculture offers a number of low-cost technologies that include:

- Altering land use.
- Changing cultivation methods.
- Better management of livestock.
- Altering crop mix and fertilization methods.
- Expanding the production of biofuels.

An example of the changes that can be made to carbon dioxide sequestration is shown in Table 3.10. A number of agricultural methods have been shown to increase the retention of carbon in soils for Canadian Prairies (Hutchinson *et al.*, 2007). Another example of changes in agricultural methods that reduce greenhouse gas emissions is the anaerobic treatment of slurry and liquid manure. Rather than placing liquid manure on the land or retaining it in lagoons where methane and nitrous oxide are produced, it can be anaerobically digested. Slurry or liquid manure anaerobically digested produces a mixture of methane (CH_4) and carbon dioxide (CO_2) which can be use as a gaseous fuel (Clemens *et al.*, 2006).

Alternative Energy Sources

The following are possible renewable, sustainable energy sources which will also reduce greenhouse emissions:

- Nuclear power.
- Hydroelectric.
- Tidal.
- Wave.
- Wind.

Table 3.10. Agricultural practices which have the potential to increase soil carbon storage. (Adapted from Hutchinson *et al.*, 2007.)

Method	Rates[a] (10^6 carbon/ha/year)	Impacts
Cropland		
Reduced tillage	0.0–0.4	Reduced erosion, enhanced biodiversity, more leaching
Eliminate summer fallow	0.0–0.5	Reduced erosion, less leaching
More forage in rotation	0.0–0.5	Improved soil structure, fertility
Increased residue return and reduction in burning	0.0–0.3	Increased production
Restore permanent grass and woodland	0.2–1.0	More wildlife
Organic residue use	0.1–0.5	–
Grazing land		
Improved practices	0.0–0.1	–
Increase in productivity (new species, irrigation, etc.)	0.0–0.3	–

[a]Rates of potential carbon gain over a projected period of 20 years.

- Geothermal.
- Solar.
- Biological.

Nuclear power

The fission process releases large amounts of energy, about 50 million times that of coal on a weight basis, which means that very little uranium fuel is required. No combustion is involved so that there are no emissions but fission generates radioactive materials, some of which have very long lives. There are also considerable problems in the reprocessing and disposal of spent fuel, the possibility of leaks or accidents, and the decommissioning of the power stations at the end of their working life. The accidents at the nuclear-generating plants at Three Mile Island and Chernobyl have shown that, despite very stringent safety arrangements, accidents can occur. This has made the public wary of nuclear power and more likely to accept alternative sources of power. Nuclear power is also regarded as not sustainable as there is a limited supply of uranium and its production involves the production of greenhouse gases.

Hydroelectric power

Hydroelectric power is a clean, non-polluting, long-lasting, renewable source, which does not produce carbon dioxide. Large-scale hydroelectric plants are responsible for about 17% of the electricity supply in developed countries and 31% in developing countries. However, hydroelectric systems have environmental impacts and can only be sited in certain areas, restricting their application.

Tidal power

The rise and fall of water level due to tides can be harnessed to generate electricity; like hydropower it is a clean, reliable and long-lasting renewable and does not produce carbon dioxide. Sites with a sufficient tidal range and area are limited and represent only 10% of the energy that is available from hydroelectricity. One of the best-known tidal power stations was built at the River Rance, Brittany, in 1966 and has been working for over 40 years producing 550 MW (Charlier, 2007).

Wave power

Schemes for the harnessing of the rise and fall of waves are under investigation in a number of countries. Devices for the conversion of wave energy to shaft power or compression have been proposed and a number have been tested.

Wind power

Harnessing the power of the wind is one of the most promising alternative methods of electricity generation as it has the potential to generate substantial amounts of energy without pollution. Wind can also be used to drive water pumps in order to store energy, to charge batteries in remote regions, or as off-grid power sources. The potential for wind power has been recognized and wind farms have been installed in at least 15 countries including Brazil, China, Denmark, Spain, USA, India and the UK (Herbert *et al.*, 2007).

Geothermal

The centre of the Earth is very hot at about 4000°C, and most of the heat which reaches the surface cannot be utilized, but in areas of volcanic activity high-grade heat is retained in molten or hot rocks at a depth of 2–10 km. The heat from these hot or molten rocks can be extracted from hot springs and used to run steam turbines directly for the generation of electricity. If the water is below 150°C, it can be used as a supply of hot water for industrial or domestic heating.

Solar energy

Sunlight can be used either directly or indirectly for solar panels for hot water generation, solar collectors for steam generation, solar architecture for heating buildings, solar thermal-electric, and steam linked to electricity generation, photovoltaic, direct generation of electricity and solar hydrogen generation. A recent review of photovoltaics was published in 2007 (Jager-Waldau, 2007).

Use of Renewable Energy

Although it is clear that energy generation from renewable resources will have to be incorporated into the overall consumption of energy, it has been slow to be adopted.

This has been in part due to the higher cost of renewable energy, reliance on the discovery of new fossil fuel sources rather than concentration on renewables, and the need for legislation to encourage these energy sources. In Chapter 1 the world's current use of energy is given, where renewables contribute almost 12.7% of the total and nuclear 6.3% (Fig. 1.4). The use of hydroelectric, biomass and other renewables is expected to rise from 1.04 Gtoe in 2002 to 1.55 Gtoe in 2030 (Table 1.2) which is actually a slight drop on a percentage basis. A more detailed breakdown of the renewables contribution is given in Table 3.11. In 2005 the world's total energy use was 11,434 Mtoe, where renewables, excluding nuclear power, contributed 3379 Mtoe.

Most of the renewable sources of energy are used to produce electricity and their contribution to world, EU (25) and UK electricity generation is given in Table 3.11. A modern power station, Didcot (in the UK), will have a peak output of 2000 MW and for nuclear station it is around 1320 MW. The peak electrical demand for England and Wales is 50,000 MW (50 GW). The potential contribution that the non-carbon-based renewable electricity generation could produce in the world is given in Table 3.12.

Table 3.11. Renewable fuels contribution to the global, EU (27) and UK electricity generation, excluding nuclear power. (From Dti, 2006a; BERR, 2007; IEA, 2007.)

Fuel	World	%	EU (25)	%	UK	%
Biomass	161	4.8	57	11.4	8.1	34.2
Waste	64	1.9	27	5.3	4.81	20.3
Hydro	2993	88.6	340	67.7	7.89	33.3
Geothermal	57	1.7	5.4	1.0	0	0
Solar PV	1.6	0.0005	1.5	0.003	0	0
Solar heat	1.1	0.0003	0	0	0	0
Wind	101	2.9	70.5	14.0	2.91	12.3
Tide	0.56	0.00017	0.53	0.001	0	0
Other sources[a]	8.8	0.003	7.0	1.4	0	0
Total	3379		502		23.7	
	(290 Mtoe)		(43.1 Mtoe)		(2.03 Mtoe)	

[a]Includes imports.

Table 3.12. Potential contribution of conventional non-carbon energy systems to electricity generation. (From Green *et al.*, 2007.)

Source	Generation (EJ/year)
World electricity generation for the year 2100 (EJ/year)[a]	146
Contribution by nuclear[b]	38
Contribution of hydroelectric	32
Contribution of solar and wind[c]	15
Electricity generation in 2100 (%)	58%

[a]If growth is 1% per year, 146 EJ/year would be required.
[b]Based on 1500, 1000 MW plants operating at 80% capacity, consuming 306,000 tonnes of uranium per year.
[c]The contribution at 50% wind and 50% solar would require 160,00 km^2 at 2116 km^2 per EJ/year (1 km^2 = 10 ha).

The possible electricity that will be required in 2100 has been included in the table to indicate the possible target. However, increased legislation and the Kyoto Protocol should encourage increases in renewable energy.

In the UK, the pattern has been much the same as the world's with nuclear and renewables producing 22.5 Mtoe in 2006 (BERR, 2007). Much of this contribution was in electricity generation. In 2005, renewable electricity generation was 4.4% (Cockroft and Kelly, 2006) not including hydroelectricity, and Table 1.4 gives the renewable energy source in terms of 1000 tonnes of oil equivalent. The major proportion comes from landfill gas and biofuels. Table 3.13 gives the peak electrical power output from some of the renewable energy systems installed in the UK at present (Dti, 2006b) and indicates some of the potentials of these systems. Small power generation systems are often connected to the distribution network rather than the main grid (400,000 volts).

Energy Storage

The demand for electricity varies daily and seasonally and therefore some centralized power stations may only be required for short periods or to operate at limited capacity. In general, a fully interconnected electricity network will use low-cost, very large power stations for the base load and more expensive units for peak loads. The fossil fuels used for either base load or peak units are easily stored and transported. Most renewable energy sources with the exception of biomass cannot be stored and transported easily, unless they are converted into electricity or other energy carriers. In addition, some renewable power systems only supply electricity intermittently, for example wind, wave and photovoltaics. In order to incorporate these intermittent electricity sources and to deal with peaks and troughs in electricity

Table 3.13. Typical peak electrical power output from renewable energy systems in the UK. (From Dti, 2006b.)

System	Peak power output MW of units	MW electricity produced	Comments
Solar photovoltaic	0.01–0.09	6	Panels are available but costly
Wind single grid connected	Up to 1	–	Some turbines are 3 MW, others 300 kW
Wind farm	10	2016	Many exist in UK but planning difficult
Geothermal	30	0	None in UK
Hydro large	Up to 130	1058	Largest in UK Loch Sloy at 130 MW average 30 MW
Tidal barrage	240	–	Severn barrage would have yielded 8640 MW, France producing 240 MW
Wave shoreline	0.18	7	Only prototypes
Wave offshore	5.25	–	Only prototypes
Burning biomass/waste	10–50	540	Size can vary
Landfill gas	1–5	781	A number of sites exist

demand some form of storing either electricity or energy which can be converted back into electricity is required. The advantages of storage would be the following:

- Bulk storage of energy would allow the decoupling of production from supply.
- Allows the incorporation of smaller power stations into the network.
- Improves power quality and reliability.
- Reduces transmission losses as transmission distances reduced.
- Cost reduction, as smaller, more efficient power stations can be constructed.
- Allows the use of intermittent renewable power sources.
- Decreased environmental impact associated with renewable sources.
- Strategic advantages of generating energy from indigenous energy sources, avoiding imports.

At present two large-scale energy storage systems are in operation: pumped hydro storage and compressed air energy storage (Dell and Rand, 2001).

In the case of pumped hydro storage, excess electricity at times of low demand is used to pump water into a lake or reservoir some distance above a hydroelectric power station. When a peak in electricity demand occurs conventional power stations are too slow to respond but the stored water can be released and the hydroelectric plant comes online rapidly. In the UK, there is such a system in Wales. The second large-scale energy storage system is to compress air in large reservoirs when electricity is in excess and release this to drive electricity-producing turbines. Such systems have been operating for some time in Germany and the USA (van der Linden, 2006).

On the small scale a number of systems are under development including the following:

- Flywheels.
- Hydrogen production.
- Batteries.
- Thermal storage.
- Superconducting magnetic coils.

Flywheels have also been used to store energy and using new technology small high-density systems have been constructed and megawatt modules can be installed.

On the island of Utsira, Norway, electricity is provided by wind turbines as there is no link to a mainland power station. Wind power is intermittent so that any excess electricity generated when the wind blows is used to electrolyse water, producing hydrogen. The hydrogen is stored and burnt to produce electricity when the wind is insufficient to run the turbines. The feasibility of a wind-photovoltaic system using compressed hydrogen has also been tested in Australia, where the costs of the hydrogen storage was the most critical factor (Shakya *et al.*, 2005).

Five types of batteries can be used to store electricity. The lead-acid battery was developed a long time ago and is used widely in the automotive industry. These batteries have also been used for small wind and solar installations but they require periodic maintenance and are poor at low and high temperatures. Alkaline batteries, nickel–iron and nickel–cadmium, were also developed a long time ago, around 1900. The best is the nickel–cadmium which performs better than the lead-acid at high and low temperatures. It is however more expensive but the nickel–metal–hydride has been developed. This battery, although more expensive, holds more charge and has seen widespread use in mobile phones and laptop computers. It has also been used in electric and hybrid vehicles. The third type of battery is the flow batteries, sometimes known as 'regenerative fuel cells' (rated to 12 MW). The cells are charged, converting electricity into chemical energy. The two compartments of the cell are separated by an ion-exchange membrane and the electrolyte in the compartments is circulated in a closed-loop system. The last two types of battery are the high temperature battery and the rechargeable lithium battery. The high temperature battery uses molten sodium at 300–400°C, and both these types have problems for large-scale use, although lithium ion batteries are widely used in portable electronic devices.

The thermal storage of energy from electricity using hot water or solid material is used to heat buildings in the form of night storage radiators where off-peak electricity is used. The heat cannot efficiently be converted back to electricity so this is not suitable for energy storage. However, phase-change materials have been used to store solar energy (Kenisarin and Mahkamov, 2007).

In systems where there is fluctuating power, superconductive magnetic energy storage can be used, and though the system is expensive, it can respond in milliseconds. Energy is stored in a magnetic field formed by a DC current in super-cooled superconductive coils.

Conclusions

The reports by Stern (2006) and the IPCC (2007) outline the consequences of global warming, and it is clear that efforts should be made to reduce the emissions of greenhouse gases from fossil fuels globally.

The IPCC has come up with four scenarios predicting the global atmospheric carbon dioxide levels depending on what measures are taken towards their reduction. If the amount of carbon dioxide released per year was retained at present values, the carbon dioxide would reach 550 ppm by 2050. Carbon dioxide emissions are still increasing so that 550 ppm may be reached before 2050. A level of 550 ppm is predicted to give a 2°C increase in global average temperature. At present there is hope to reduce carbon dioxide release, so that a value of 450 ppm is reached by 2100. Small increases in temperature seem insignificant, but these can have far-reaching effects such as the melting of sea ice. Some consider that even if we stopped carbon dioxide emissions now, the tipping point may have already been reached and a rapid and long-lasting increase in temperature is inevitable.

The chapter outlines the methods that are currently available or under development for the reduction of atmospheric carbon dioxide which include: burn less fuel, sequester the carbon dioxide and use non-carbon-dioxide-producing energy. Within these broad categories there are many options and no one option will provide a complete solution, but in concert they may well affect the outcome. The solution is not the science but rather the politics where countries have to reduce carbon dioxide emissions, while at the same time producing growth in their economies and increasing prosperity. It seems that if energy supply is to be sustainable and carbon-neutral it cannot be obtained at the same time as continued growth of the economy. However, developing countries will not stop their development in order to reduce carbon dioxide emissions despite the warning that global warming will affect developing countries the most. Considerable political effort and legislation will be needed if global warming is to be halted.

4   Biological Solid Fuels

The Nature of Biofuels: First-, Second- and Third-generation Biofuels

The alternative energy sources are derived from biological material and it is these sources that are the main focus of the book. Recently the use of biological materials to provide a source of energy that is renewable and can mitigate carbon dioxide accumulation has attracted considerable attention (Chum and Overend, 2001; Hamelinck *et al.*, 2004; Cockroft and Kelly, 2006).

The range of biofuels that can be produced is listed below and includes biofuels that are being used at present and others which are still at the development stage. Biological-based fuels can be solid, liquid and gaseous, and the physical state of the fuel greatly influences the way it is used. It is the developmental stage that has been used to divide biofuels into first-, second- and third-generation biofuels (Fig. 4.1). Those biofuels currently used and produced in large quantities are the first-generation biofuels. The biofuels that have been produced but technical difficulties and high costs have delayed their application on a large scale are the second generation. The third-generation biofuels are those which are still at the research and development stage.

Solid fuels:

- Biomass.
- Wastes.

Gaseous fuels:

- Methane (biogas).
- Hydrogen.
- Dimethyl ether (DME).

Liquid fuels:

- Methanol (FT origin).
- Ethanol.
- Biobutanol.
- Synthetic petrol (FT origin).
- Synthetic diesel (FT origin).
- Biodiesel (esters).
- Biodiesel (bio-oil).
- Biodiesel (plant and microalgal hydrocarbons).
- Biodiesel (microalgal oils).

The first-generation biofuels are represented by biomass, biogas, biodiesel and ethanol. Biomass is not included in Fig. 4.1 as it is mainly combusted or co-fired with coal to produce electricity. However, biomass in the form of wastes and lignocellulose can

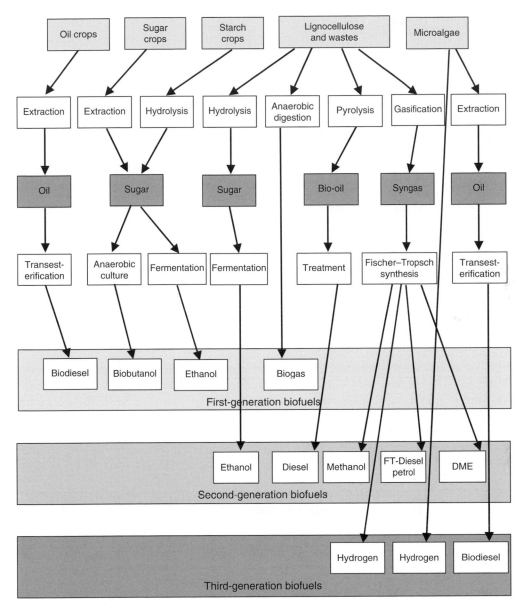

Fig. 4.1. The sources and processes for the production of first-, second- and third-generation biofuels.

also be converted into biogas, diesel, petrol, methanol and dimethyl ether using gasification (Fig. 4.1).

The first-generation biofuels are produced from energy crops such sugarcane, sugarbeet, maize, wheat, rapeseed, soybean and sunflower. However, to completely replace the fossil fuels gas, petrol and diesel large areas of land will be required. Hence, there is not enough land to grow sufficient energy crops without competing with food crops for land. For these reasons second- and third-generation biofuels are under development (Fig. 4.1). The second-generation biofuels will be produced from ligno-cellulose biomass and wastes which have much better yields per hectare as the whole of the harvested plant will be used. The higher yields will mean that second-generation biofuel production will compete less with food crops. The direct production of hydrogen and extraction of oil for biodiesel from microalgae are third-generation biofuels which will not compete with food crops. Microalgae can be grown on non-agricultural land or in marine conditions and because they are some 50–100 times more productive than biofuel crops much less land will be required (Chisti, 2007).

Therefore, it is essential that second- and third-generation biofuels are developed as first-generation biofuels can only realistically supply 5% of the fuels required.

Introduction to Solid Biofuels

This chapter deals with the potential, the use and the future of solid biofuels.

Biological material in the form of wood, specific crops, crop residues and organic wastes fulfil both renewable and sustainable criteria as forests and crops can be replanted and, provided these are renewed, they are sustainable. These materials are often referred to as biomass which has been defined as the yield of organic matter that may be used as a source of energy and/or chemicals. Since biomass is largely of plant origin, it can be more correctly referred to as phytomass, although biomass is the widely accepted term. When biomass is burnt it can be regarded as carbon dioxide neutral as any carbon dioxide released during their combustion has previously been fixed from the atmosphere during photosynthesis. However, in the cultivation, harvesting, preparation, transportation and processing of biomass, fossil fuels may be required, thus making them less than 100% carbon neutral. The world's total annual energy use has been estimated to be 425 EJ and estimates of the contribution that biomass could make vary from 7.3 to 15.0% of this total (Boyle, 1996; Venturi and Venturi, 2003; Parikka, 2004; IEA, 2005a; Faaij, 2006) (Table 4.1).

Table 4.1. Population and energy consumption of biomass in small and large countries. (Adapted from Wright, 2006; the data is for 2002.)

Country	Population (millions)	Energy use total (EJ)	Biomass energy (EJ)	Biomass energy (%)	Energy crop used
China	1295	45.5	7.5	16.4	Fuelwood 70,000–100,000 km^2
EU 25	453	70.5	2.75	3.9	District heat 180 km^2 willow and grasses
USA	288	103.4	2.92	2.8	500 m^2 woody crops
Brazil	177	7.3	1.98	27.2	Charcoal from 30,000 km^2 Ethanol from sugarcane
Canada	31	13.1	177	13.5	Forest residues
Australia	20	5.2	0.2	3.8	
UK	59.7	9.48	0.06	0.6	25 km^2 willow
Sweden	8.9	2.2	0.34	15.9	160 km^2 willow and reed canary grass
Netherlands	16.1	3.6	0.083	2.3	1.2 km^2 willow and grasses
Denmark	5.4	0.83	0.098	11.8	Small-scale trials

Parikka (2004) estimates that the global use of woodfuel and firewood was 3271 Gm³/year (39.7 EJ) of which 55% is used a fuel (21.8 EJ) and 45% as round-wood (17.9 EJ). As can be seen the less developed countries obtain a higher proportion of their energy from biomass (16.4–27.2%), as do Sweden and Denmark where efforts have been made to exploit their forest resources.

The Types of Biomass Available

There are five categories of solid biofuels, depending on their source. The yield and energy content of some of the biomass types are given in Table 4.2.

1. Wood from forests and forest residues.
2. Crop residues.
3. Crop specifically grown for energy generation.
4. Animal waste.
5. Municipal waste.

Wood biomass

Materials, such as wood, have always been used by humans as a source of energy, but depending on the development of a country the amount of biomass used can differ greatly (Table 4.1).

The wood from forests consists of felled wood, thinnings, logging residues, wood-processing residues and waste from clearing. The wood available will be that surplus to the need for construction and industrial wood products from forests, plantations and trees outside forests.

Table 4.2. The direct use of biomass, wood, straw and SRC for the production of electricity.

Biomass	Fuel type	Species	Yield (tonnes dry weight/ ha/year)	Energy content GJ/tonne
Woodland/forests	Chips	Many	10–35	15
Specific energy crops				
Short rotation coppice	Chips	Willow	6–15	15
(SRC)	Chips	Poplar	10–17	15
Perennial grasses	Chips	Miscanthus	20	15
	Chips	Limpograss	7–22	15
	Chips	Napier grass	34–55	15
Crop residues	Waste	Bagasse	20	14
		Maize straw	20	14
		Straw from wheat,		
		barley, millet	20	14

Crop residues

Crop residues include haulms of legumes, stalks of sorghum, maize and millet and straw from rice, wheat and barley. Although there is a considerable quantity of biomass as residues, it is widely distributed and seasonal in its availability. The yields can be up to 20 t/ha with an energy content of 14 GJ/t giving a yield of 280 GJ/ha.

Specially grown energy crops

Crops specifically grown for energy biomass include short rotation coppice with willow (*Salix* sp.) and poplar (*Populus* sp.), perennial grasses miscanthus, switchgrass (*Panicum virgatum*), reed canary grass (*Phalaris arundinacea*) and giant reed (*Arundo donax*).

Coppice has been used in Europe for centuries with long rotations of 10–30 years using hazel (*Corylus avellana* L), black alder (*Alnus blutinosa* L), and to a lesser extent oak (*Quercus robur* L), ash (*Fraxinus excelsior* L), elm (*Ulmus* sp.) and hornbeam (*Carpinus betulus* L). Latterly non-indigenous species have been used. In the coppice system the stems are cut back at close to soil level after 3–10 years. This encourages the growth of a large number of stems which can be harvested after 3–4 years in the case of willow and poplar. The other species have a slower growth rate and will be harvested after 10–30 years. Table 4.2 gives the yield and energy content of the short rotation coppice species. Willow and poplar yield between 6 and 17 t/ha/year with a calorific value of 15 GJ/t. Plantations of 15,000 cuttings/ha for willow and 10–12,000 cuttings/ha for poplar are planted in winter or spring. After 1 year the single stem is cut back to soil level to produce a multi-stemmed stool which can be harvested every 3–4 years. Harvest is mechanical and the stems chipped and dried for combustion. Willow is the preferred species in the UK, where the plantations can last for 25–30 years. Much of the biomass is co-fired with coal in power stations.

Perennial grasses

Perennial grasses have in the past been used as fodder crops but now they are considered suitable as energy crops because of their high content of cellulose and lignin. This gives the plant biomass a high heating value. Some examples of the heating values of biomass types are compared with fossil fuels in Fig. 4.2. The figure shows that the energy density of all the biomass types is lower than coal and especially gas and oil. As a consequence more biomass will be required to produce an equivalent amount of energy and thus more biomass will need to be transported.

Trials with a large number of perennial grasses have been carried out for energy in both the USA and Europe (Table 4.3). The criteria that were used in the selection as an energy crop were as follows:

1. Suitable for the climate in the region.
2. Easily propagated.
3. A consistent and high yield of biomass per hectare, probably the most important.
4. Positive balance of energy input versus output.
5. The crop can be cultivated in a sustainable manner.
6. Resistance to pests and diseases.

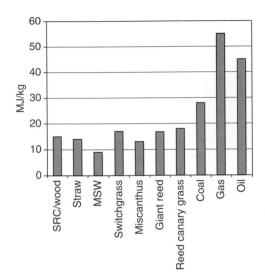

Fig. 4.2. Energy content of fossil fuels, SRC and perennial grasses.

Table 4.3. Perennial grass species tested in the EU as an energy crop. (From Lewandowski *et al.*, 2003.)

Name	Latin name	Photosynthetic pathway	Yields (t/ha/year)
Meadow foxtail	*Alopecurus pratensis*	C3	6–13
Big bluestem	*Andropogon gerardii*	C4	8–15
Giant reed	*Arundo donax*	C3	3–37
Cypergras, Galingale	*Cyperus longus*	C4	4–19
Cocksfoot grass	*Dactylis glomerata*	C3	8–10
Tall fescue	*Festuca arundinacea*	C3	8–14
Raygras, ryegrass	*Lolium* sp.	C3	9–12
Miscanthus	*Miscanthus* sp.	C4	5–44
Switchgrass	*Panicum virgatum*	C4	5–23
Napier grass	*Pennisetum purpureum*	C4	27
Reed canary grass	*Phalaris arundinacea*	C3	7–13
Timothy	*Phleum pratense*	C3	9–18
Common reed	*Phragmites communis*	C3	9–13
Energy cane	*Saccharum officinarum*	C4	27
Giant cordgrass/salt reedgrass	*Spartina cynosuroides*	C4	9/5–20
Prairie cordgrass	*Spartina pectinata*	C4	4–18

7. Broad genetic diversity to enable species to be adapted to prevailing conditions.
8. Harvesting possible with existing technology.
9. Perennial.
10. Competitive on cost with food crops.

The four that have been chosen for further study are Miscanthus, switchgrass, reed canary grass and giant reed. Some of the properties of these grasses are compared with short rotation coppice of willow and poplar in Table 4.4.

Table 4.4. Properties of biomass crops. (From Powlson *et al.*, 2005; Lewandowski *et al.*, 2003.)

Crop	Poplar (SRC)	Willow (SRC)	*Miscanthus* sp.	Switchgrass	Reed canary grass	Giant reed
Yield t/ha/year	7	7 (15–30)	12 (5–44) (10–25)	10 (5–23) (15–35)	8 (7–13)	5–23
Establishment time	3 years+	3 years+	3 years+	2–3 years+	1–2 years	1–2
Photosynthetic pathway	C3	C3	C4	C4	C3	C3
Fertilizer	Low/medium	Low/medium	Low	Very low	Medium	Moderate
Water supply	Wet	Wet	Not tolerant to stagnant water	Drought tolerant	Drought tolerant	Drought tolerant
Pesticide	Low	Low	Low	Very low	Low	Low
Establishment costs	High	High	Very high	Very low	Very low	High
Pest/disease	–	Beetle rust	None	None	Some insect problems	Few
Day/length	Long	Long	Long	Short	Long	Long
Plantation longevity	20 years	20 years	20 years	20 years	10 years	n/a
Energy content GJ/t	15	15	17.6–17.7	17.4	16.5–17.4	17.3–18.8
Output GJ/ha/year	105	105	260–530 (262–525)	174–435 (175–437)	240–600 (262–613)	88–403

Miscanthus

In Europe research on perennial grasses started with Miscanthus. The genus *Miscanthus* contains 17 species and originates from East Asia and a hybrid *Miscanthus × giganteus* was first introduced into Europe in 1930 from Japan as an ornamental. Miscanthus grows vigorously and can be harvested dry in one harvest. Miscanthus is a C4 metabolism grass which can reach 4 m in height and forms rhizomes. Different Miscanthus species have different rhizomes, which are persistent, with the oldest plantation some 18 years old. Miscanthus is wind pollinated with fan-shaped inflorescences. Miscanthus can be grown on a wide range of soils but does not tolerate waterlogged soils. Most of the yields for Miscanthus reported in Europe have been determined using *Miscanthus × giganteus*. Yields are variable with values in the range of 5–44 t/ha/year. It does however suffer from some problems of poor resistance to cold and high costs of propagation as rhizomes have to be used.

Switchgrass

Switchgrass is a native of the North American grasslands, a perennial C4 grass with a high yields on poor soils. Like Miscanthus it was introduced into Europe as an ornamental grass, but based on the data obtained in the USA it has been considered as an energy crop. It is perhaps the best choice as it is drought-tolerant, gives high yields and can be harvested once a year.

Reed canary grass

Reed canary grass is a species indigenous to Europe belonging to the Gramineae family. It is adapted to low temperatures and short growing times. Reed canary grass is a C3 grass which grows to 3 m in height, is propagated by seed and is harvested once a year.

Giant reed

Giant reed is also an indigenous species belonging to the Gramineae family. It is a tall perennial C3 grass which can reach heights of up to 8–9 m. The giant reed tolerates a variety of conditions but prefers well-drained soils. It is propagated by rhizomes rather than seed and the yield can reach 100 t/ha/year under optimum conditions.

All perennial grasses are regarded as drought tolerant, require few inputs and grow on poor land. However, establishment of these crops is not easy and can vary greatly. Yields of biomass can also vary and appear to be related to nitrogen and water availability. Switchgrass requires as much water as traditional crops and is responsive to nitrogen fertilizer but too much usage will give a problem of lodging.

C3 and C4 metabolism

One of the important characteristics of some of the perennial grasses is the possession of C4 metabolism rather than C3. C4 and C3 metabolism refer to the pathways used to assimilate carbon dioxide during photosynthesis. C4 plants are more efficient at higher light and temperatures compared to C3 plants. The C4 plants have a lower moisture content, require less fertilizer input and are twice as efficient with water. C4 assimilation of carbon is theoretically 350 kg/ha/day compared with 200 kg/ha/day for C3 plants (Venturi and Venturi, 2003). All these features make C4 plants more suitable for biomass fuel planting than C3 plants.

The development of the C4 metabolism of carbon dioxide assimilation evolved from the Calvin cycle in C3 plants to avoid the loss of carbon dioxide through photorespiration. The fixation of carbon dioxide during photosynthesis takes place in three stages. The addition (carboxylation) of carbon dioxide to ribulose-1,5-bisphosphate (Fig. 4.3) is followed by the reduction of 3-phosphoglycerate to glyceraldehyde-3-phosphate. Ribulose-1,5-bisphosphate is then regenerated from glyceraldehyde-3-phosphate. The first step of the Calvin cycle is catalysed by the chloroplast enzyme ribulose bisphosphate carboxylase/oxygenase known as rubisco. However, another property of the rubisco enzyme is to catalyse the oxygenation of ribulose-1,5-bisphosphate which is the start of light-dependant oxygen uptake and carbon dioxide release, known as photorespiration, which reduces plant yield.

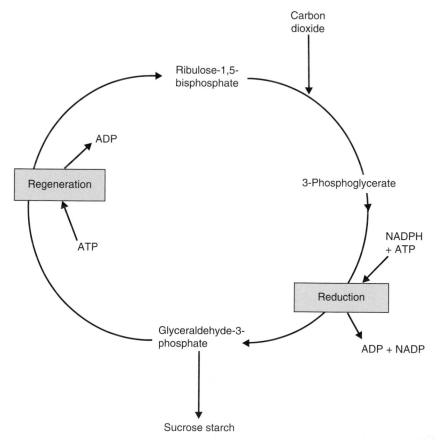

Fig. 4.3. The Calvin cycle involves the fixation of carbon dioxide through photosynthesis (C3 metabolism).

In the C4 metabolism two different types of cell are involved in photosynthesis, the mesophyll and bundle sheath cells (Fig. 4.4). In the mesophyll cells carbon dioxide is used to carboxylate phosphoenolpyruvate (PEP) forming oxaloacetate. The oxaloacetate is converted to malate (C4) and this is transferred to the bundle sheath cell where the malate is converted into pyruvate and carbon dioxide. The carbon dioxide is then used in the Calvin cycle. The C4 cycle has a higher energy demand but the cycle reduces photorespiration and water loss. The phosphoenolpyruvate (PEP) carboxylase enzyme has a high affinity for the carboxyl ion such that it is saturated and in equilibrium with carbon dioxide gas. Oxygen is not a competitor in the PEP

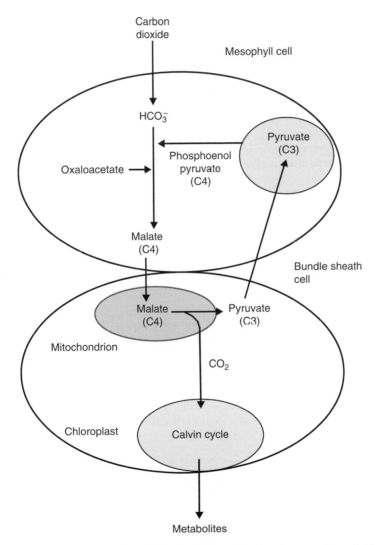

Fig. 4.4. C4 metabolism in plants involves two types of cells where carbon dioxide is used to carboxylate phosphoenol pyruvate forming oxaloacetate, which is converted into malate. The malate is transferred to the bundle sheath cells where it is split into pyruvate and carbon dioxide. The carbon dioxide is used in the Calvin cycle.

carboxylase reaction because the substrate is a carboxyl ion. The high activity of the PEP carboxylase allows the plants to reduce the stomatal opening, reducing water loss, while fixing carbon dioxide at an undiminished rate. The high concentration of carbon dioxide in the bundle sheath cells allows the cells to carry out photosynthesis at high temperatures.

Animal and municipal waste

Animal wastes consist of excess slurry and dung from cattle, chickens and pigs. These wastes can be used to generate biogas through anaerobic digestion and there are a number of farm-sized units available. There are also a number of small electricity power stations which run on chicken slurry from large battery chicken farms.

Potential Biomass Available

Worldwide sources of biomass

One of the problems with biomass is the question of whether there is sufficient biomass to replace fossil fuels and can it be transported to where it is used. It has been estimated that biomass contributes 39.7–45 EJ/year to the global energy supply (9–15%) from a total 425 EJ (Parikka, 2004; Faaij, 2006). There have been a number of estimates of the biomass available globally and the range of estimates is given in Table 4.5. These estimates vary considerably from 1135 to 93 EJ. A potential of 100 EJ would supply around 25% of the present global energy requirement. Smeets and Faaij (2007) estimate that there may be a surplus of biomass, after directly used woodfuel and roundwood have been utilized, of 71 EJ but this may be an underestimate. Considering ecological and technical considerations the biomass may be only 2.4 Gm^3 (28 EJ). The potential global biomass energy contained in various biomass types is given in Table 4.6. The great variation noted by Hoogwijk *et al.* (2003) appears to be in the estimate of the biomass grown on surplus or marginal land. Marginal lands are those with little economic value but before these can be used the environmental impacts need to be determined. The marginal land may be a specific wildlife habitat such as a wetland or forest.

The potential biomass energy consisting of wood, energy crops and straw in the various regions is given in Table 4.7, where the total is given as 103.8 EJ. All regions except Asia use only a small proportion of the available energy.

Table 4.5. Estimates of the potential global biomass available.

Biomass energy potential in EJ	Reference
1135	Berndes *et al.* (2003)
110	Hoogwijk *et al.* (2003)
114	Fischer and Schrattenholzer (2001)
93	Dessus *et al.* (1992)
103.8	Parikka (2004)

Table 4.6. Global biomass energy potential from various sources. (From Hoogwijk *et al.*, 2003.)

Source	Comments	Potential energy EJ/year
Biomass on surplus land	Area 0–2.6 Gha	0–988
Biomass on degraded land	430–580 Mha	8–10
Agricultural residues	Various estimates	10–32
Forest residues	Sustainable potential from various studies	10–16
Animal manure	Various estimates	9–25
Organic wastes	Various wastes	1–3
Bio-materials	Replacements for chemicals, plastic, paints and solvents	83–116
Total		33–1130

Table 4.7. Potential biomass energy in world regions in EJ (exajoules, 10^{18}). (From Parikka, 2004.)

Biomass	North America	South America	Asia	Africa	Europe	Middle East	USSR	World
Wood	12.8	5.9	7.7	5.4	4.0	0.4	5.4	41.6
Energy crops	4.1	12.1	1.1	13.9	2.6	0	3.6	37.4
Straw	2.2	1.7	9.9	0.9	1.6	0.2	0.7	17.2
Other	0.8	1.8	2.9	1.2	0.7	0.7	0.3	7.6
Potential	19.9	21.5	21.4	21.4	8.9	0.7	10.0	103.8
Use EJ/area	3.1	2.6	23.2	8.3	2.0	0	0.5	39.7
Potential used (%)	16	12	108	39	22	7	5	38

Potential biomass use in the UK

In the UK, the total energy consumption in 2006 was 232 Mtoe with the consumption of gas at 89.2 Mtoe and coal 43.4 Mtoe (Fig. 1.4). Both gas and coal are used to generate electricity and it is these two fuels that biomass may replace. An estimate of the possible biomass available in the UK is given in Table 4.8. A total of 7 million t of biomass represents an energy content of 0.149 EJ as 1 t of biomass represents 19 GJ. This constitutes 1.53% of the total energy consumed in the UK. If the biomass is converted into electricity at an efficiency of 30%, it would generate 9.8 TWh. The UK electricity use is 346.4 TWh and therefore 9.8 TWh represents 2.83% of total demand.

In another study the potential of a mixture of energy crops and forestry and straw wastes combined with a conversion of grassland to produce electricity was determined (Powlson *et al.*, 2005) (Table 4.9). The mixture of energy sources was capable of producing 12.2% of the electricity required from 7.43 Mha, some 40% of agricultural land. In contrast it has been suggested that 2.7 Mha would be required to produce 100% of the UK's electricity (Rowe *et al.*, 2009).

Table 4.8. The available biomass in tonnes dry material in the UK. (From Woodfuel, 2007; www.woodfuel.org.uk)

Source	England	Scotland	Wales	Total
Forest and woodland	2,394,147	2,942,513	971,689	6,308,349
Thinnings	616,060	34,717	19,706	670,483
Short rotation coppice	15,899	572	218	16,689
Waste	289,686	403,538	165,783	858,901
Total	3,315,686	3,381,340	1,157,396	7,854,422

Table 4.9. Potential electricity from biomass. (Adapted from Powlson *et al.*, 2005.)

Process	UK electricity (%)	GWh
80% set-aside (611,000 ha in 2002)	2.7	9,282
50% sugarbeet converted to biomass (169,000 ha)	0.5	1,719
50% forestry waste	1.6	5,501
50% wheat straw (19 Mt, thus 8 Mt from 2 Mha)	3.7	12,720
10% grassland converted to biomass 6.65 Mha	3.7	12,720
Total	12.2	41,943

Electricity demand 343.8 TWh (343,800 GWh), 1.237 EJ. Electricity generation at 1.6 MWh/t, 12 t/ha and efficiency of 30%.

Energy and Fuel Generation Using Biomass

There are four processes whereby biomass can be used to generate electricity or produce fuels. These options are given in Fig. 4.5 and include direct combustion, co-firing, gasification and pyrolysis.

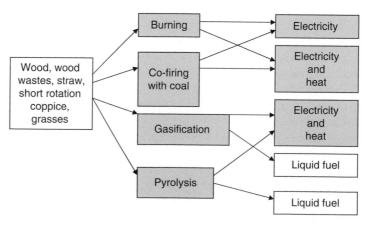

Fig. 4.5. The direct use of biomass for the production of fuel and energy. Heat and electricity is produced through combined heat and power systems (CHP).

Combustion

Biomass has been used for many years to provide domestic heating using direct combustion in fires and stoves. These systems are normally not very efficient and produce some emissions of soot and dust. Small to medium-scale heating systems using pelleted biomass have been developed for houses and larger buildings such as schools. These modern heating systems have improved efficiencies and reduced emissions. Large-scale combustion of biomass for the production of electricity is used in many countries. The technology involved in the combustion of wood and forest residues can be conventional pile burning, stationary, moving, vibrating, suspension and fluidized beds. Wastes can be incinerated to produce heat and power and are found widely distributed in Europe. Incineration is central to the treatment of domestic waste although there have been some concerns about emission from incinerators. The application of fluidized beds and advanced gas cleaning has given an efficiency of 30–40% for electricity production at a scale of 50–80 MW combined with flue gas cleaning.

Combined heat and power (CHP)

The application of combined heat and power (CHP) run on biomass for district heating has been applied widely in Scandinavia and Austria. CHP has the advantages of higher efficiency of electricity generation and lower costs, and these systems have also encouraged the development of a biomass market. The burning of straw for CHP has seen the development of more complex boilers and pre-treatment of the straw. An example of a system using biomass to produce electricity is shown in Fig. 4.6. Short rotation coppice or perennial grasses are chipped or made into pellets and dried using some of the waste heat from the turbine before use. The flue gas can be treated before release and the only waste that needs disposal is the slag from the boiler.

Co-firing

The addition of biomass to coal for use in power stations occurs in the UK and many European countries. The advantages are that overall emissions of GHGs are reduced by incorporating biomass, the efficiencies are high due to the large scale, and investment costs are low making this an attractive GHG mitigation option. Co-firing with coal is used for the generation of electricity and has been shown to reduce emissions of sulfur dioxide and NO_x. Normally the co-firing proportion is around 10% which changes the processes very little. The biomass that has been used includes straw, short rotation coppice (SRC), sawdust and other wastes such as sewage sludge, manure and municipal solid waste (Sami *et al.*, 2001).

Gasification

Biomass can be converted into gas by heating at 1300°C in an oxygen-limited atmosphere. The gas produced contains mainly hydrogen, carbon monoxide, methane, carbon dioxide and nitrogen and can be used in a boiler or turbine for the generation

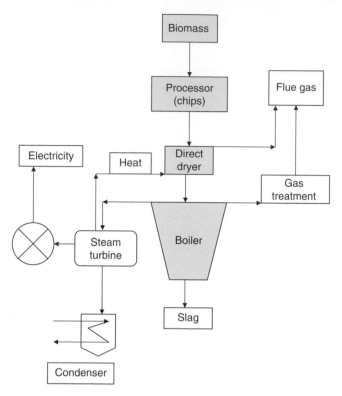

Fig. 4.6. The direct use of biomass, wood, straw and short rotation coppice (SRC) for the production of electricity.

of heat, steam or electricity. Small-scale gasification systems (up to 100 MWth; megawatts thermal) have been developed for heat and power systems. Small fixed-bed gasifiers linked to diesel or gas engines (100–200 kW) have an electrical efficiency of 15–20% but as yet have not been installed. The variations in the biomass used for gasification makes the small systems difficult to operate. Larger gasifiers over 100 MWth use a variety of systems, including circulating fluidized beds, atmospheric gasifiers (ACFB), integrated gasification/combined cycle (IGCC) (Faaij, 2006). An example of simple-cycle gas turbine and biomass integrated gasification combined cycle (BIGCC) is shown in Fig. 4.7. The efficiency of the simple-cycle gas turbine can be improved from 40 to 60% with a combined cycle turbine system (Rukes and Taud, 2004).

Pyrolysis

Pyrolysis is the heating of the biomass in the absence of air at temperatures of 300–500°C. Under these conditions the products are gas, charcoal and an oil (bio-oil) which after treatment can be used in a diesel engine. The main treatment is to reduce the viscosity which is too high to be used in a diesel engine. The possible uses of pyrolysis products are given in Fig. 4.8. The crude bio-oil can be used in gas turbines and engines but for the standard diesel engine it requires upgrading. The bio-oil can also be used in boilers and co-fired in power stations and after gasification it can be

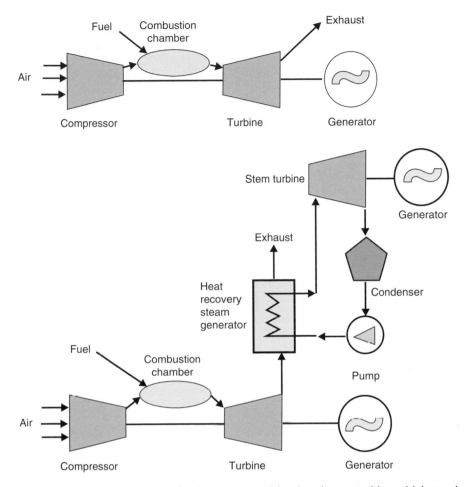

Fig. 4.7. Simple-cycle gas turbine (top) and the combined cycle gas turbine which can be used with gasified biomass (biomass integrated gasification combined cycle, BIGCC).

converted into transport fuels. It can be a source of chemicals. The charcoal can be used for industrial processes or as a source of heat. The pyrolysis liquid can have a number of names such as pyrolysis oil, bio-oil, bio-crude-oil and wood oil. Bio-oil is a dark brown acidic liquid consisting of a complex mixture of oxygenated hydrocarbons and water which is not miscible with petroleum-based fuels. Some of the properties of wood-derived bio-oil are given in Table 4.10.

Bio-oil can replace fuel oil in static operations such as boilers, furnaces, engines and turbines for the production of heat and electricity but to be used as a transport fuel the viscosity needs to be reduced.

Conclusions

Biomass is a solid biofuel which can be in all forms of wood – from trees, crop residues, animal and municipal waste – in addition to crops specifically grown for energy.

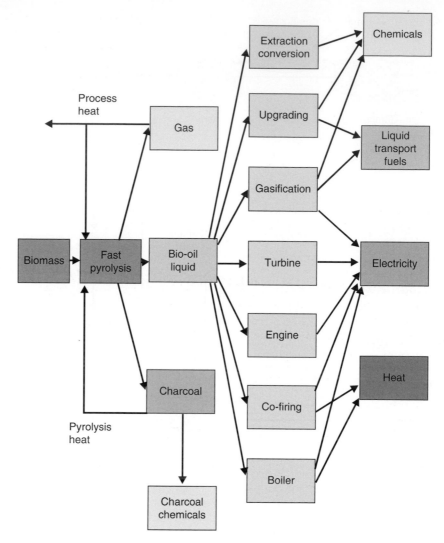

Fig. 4.8. Possible application for bio-oil obtained by the pyrolysis of biomass.

Table 4.10. Properties of wood derived bio-oil. (From IEA Bioenergy update 29, 2008a.)

Property	Value	Comments
Moisture	25%	From moisture in feed
pH	2.5	From organic acids
Density	1.2	Dense compared with other fuels
Elemental analysis	C 57%; H 6%; O 37%; N trace	
Ash	0%	Stays with char
Viscosity (40°C, 25% water)	50 cp	Can vary from 20 to 1,000 cSt
Solids	0.2%	Char
High heating value	18 MJ/kg	

Energy can be extracted from biomass by direct burning, co-firing with coal, gasification and pyrolysis. Gasification and pyrolysis can yield liquid fuels and can be used in electricity generation. The process for the generation of petrol and diesel in the Fischer–Tropsch (FT) process is discussed in Chapter 7. The main use of biomass has been in the generation of electricity and in combined heat and power systems.

The estimates of the energy available in biomass vary greatly from 93 to 1135 EJ against the global energy requirement of 425 EJ. It is clear at first glance that biomass can provide a significant amount of energy and Table 4.7 indicates that most regions do not use a high proportion of their potential. Biomass use is perhaps more efficient than biodiesel and bioethanol as the whole plant can be used rather than only a small proportion. However, other studies indicate that there may have been an overestimation of the biomass actually available. The surplus biomass may be much smaller after woodfuel has been taken into consideration. In addition, biomass is widely spread and seasonal and will require collection and transport if it is to be used and this will reduce the amount available. The other problem is that biomass can only be used once and there are insufficient amounts available for all uses especially in the UK. It may be that the FT process for diesel and petrol production will be the most suitable for biomass. For biomass to be adopted as a sustainable source of energy, it requires government help in terms of tax concessions and provision of sites. The lack of government support is not the only impediment to the introduction of biofuels, as local opposition to renewable schemes such as wind farms has been strong in many cases. One example of local opposition triumphing over government initiative was the projected biomass electricity generation plant in Cricklade, Wiltshire, which was not granted planning permission (Upreti and van der Horst, 2004). A number of other similar schemes (27%) have been rejected for similar reasons. The plan was to build a 5.5 MW power station at Kingshill Farm, an established recycling site, to generate electricity under a Non-fossil Fuel Obligation (NFFO 3) contract capable of supplying electricity to more than 10,000 homes, at a time when Swindon was expanding rapidly. The site required 36,000 t of dry wood supplied from a 30-mile radius using forestry wastes and SRC. The rationale for the site was as follows:

1. Good access to forestry wastes.
2. The area was suitable for growing SRC.
3. Good road connections for fuel delivery.
4. Good access to electricity distribution.
5. It delivered electricity in a decentralized location.
6. Local employment, 15 permanent jobs and other jobs during construction.
7. Diversification in local agriculture.

However, despite these points individuals and organizations opposed the development and the main objections included:

- It would set a precedent for other local industrial developments.
- It would contradict local designations, the Area of Special Archaeological Significance and Rural Buffer Zone.
- It would lead to a large increase in the movement of heavy goods vehicles (HGVs).
- The six chimneys proposed were very tall and would affect the view.

- Approximately 117 million l of water would be lost into the atmosphere.
- The power station would produce smell, dust, noise and other emissions.
- Long-term general health impacts.
- Damage to Cricklade's south-east meadows and water systems.
- Possible lack of compensation if anything should go wrong.
- There would be a negative effect on property prices.

These concerns were clearly very powerful and in September 2000 the application was rejected for the following reasons:

> The Biomass Power Station is a major development proposal which would, if allowed, seriously undermine the openness of the rural landscape, resulting in a loss of countryside creating an inappropriate form of major development in the Rural Buffer, contrary to the Wiltshire Plan Review and Policy DP 13 of the Wiltshire County Structure Plan 2011 Proposed Modifications.
>
> (Upreti and van der Horst, 2004)

It was concluded that public relations strategies by developers, role of the media in amplifying risk, lack of proper information and lack of public understanding of biomass power plants were the main reasons for the lack of success. The UK government needs to generate a public awareness of the benefits and need for sustainable electricity generation or NIMBYism will prevail.

5   Gaseous Biofuels

Introduction

In contrast to the solid biofuels, described in Chapter 4, gaseous biofuels can not only be used for both electricity generation and heating, but also most importantly as a transport fuel. A list of gaseous biofuels is given below:

Gaseous fuels:

- Methane (biogas).
- Hydrogen.
- Dimethyl ether (DME).

Methane or biogas can be used to replace natural gas (methane) which is a fossil fuel for electricity generation and for cooking and heating. For land transport, there are a small number of modified internal combustion engines using gases derived from fossil fuels such as liquid natural gas (LNG), liquid petroleum gas (LPG) and compressed natural gas (CNG). Biogas, hydrogen and dimethyl ether have been proposed as replacements for these transport fuels. Hydrogen has also been proposed as a fuel for gas turbines.

Gaseous fuels have problems of storage and supply not encountered with either solid or liquid fuels. Storage of gas at atmospheric pressure is not practical so the gas has to be compressed to high pressure or liquefied at low temperatures to reduce its volume. Compression to pressures of 200 bar and liquefaction, which for hydrogen needs a temperature of –253°C, expends a considerable amount of energy and subsequent storage has to be in strong pressure vessels or in well-insulated tanks. The lower energy density of the gaseous fuels compared with liquid fuels means that larger fuel tanks are required in vehicles. One advantage is that transport of gaseous fuels can be carried out using pipelines which are used at present for natural gas although in the case of hydrogen its low density may encourage leaks. All gaseous fuels are inflammable, especially hydrogen, which introduces safety problems when these gases are stored in vehicles. The dangers of hydrogen fires are often illustrated by the crash of the airship *Hindenberg*, but as hydrogen diffuses so rapidly any spill in an open space may disperse before anything can happen.

Methane (Synthetic Natural Gas and Biogas)

Methane (CH_4) is a natural gas produced by the breakdown of organic material in the absence of oxygen in wetlands, termite mounds, and by some animals. In addition, methane is a greenhouse gas which is 23 times as effective as carbon dioxide, but because of its low concentration is only responsible for 15% of global warming. Mankind is also responsible for the release of methane through biomass burning, agriculture (rice paddies), cattle, and release from gas exploration (Chapter 2, Table 2.5).

The reasons for considering methane (biogas) as a possible biofuel are as follows:

- Increases in the costs of waste disposal due to regulation and taxes have encouraged the investigation of alternative methods of disposing of waste.
- The EU directive on use of renewable fuels and the Renewable Transport Fuel Obligation (RTFO) in the UK, and methane is a renewable fuel.
- The greater use of biomass in the UK.
- Improvements in air quality by the introduction of biofuels.
- Reduction in methane released into the atmosphere to comply with the Kyoto Protocol.
- Reduction in natural gas imports as much of the natural gas is supplied from unstable areas.

Methane is produced under anaerobic (no oxygen) conditions where organic material is broken down by a consortium of microorganisms. The three main sources of material for anaerobic digestion are given in Fig. 5.1.

In the UK, biogas is produced by anaerobic digestion of sewage sludge (190,000 t of oil equivalent) but landfill produces the bulk of the gas (1,320,000 t of oil equivalent) (Fig. 5.2). The bulk of the biogas is used to generate electricity. It has been estimated that the UK is capable of producing 6.3 million t of oil equivalent (Mtoe) as methane (NSCA, 2006).

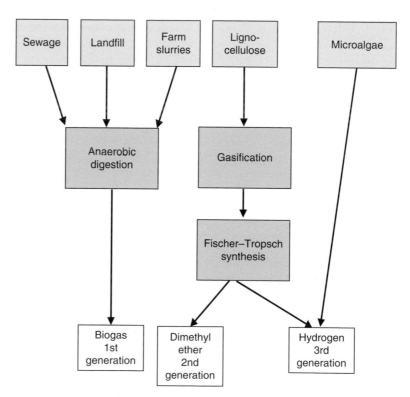

Fig. 5.1. Production of gaseous biofuels which are first, second and third generation.

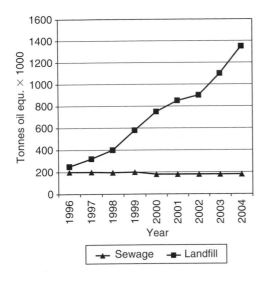

Fig. 5.2. Biogas (methane) in tonnes of oil equivalents produced in the UK from anaerobic digestion of sewage and managed landfill sites. (Redrawn from NSCA, 2006.)

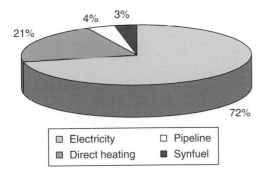

Fig. 5.3. Uses of landfill biogas in the USA. (Redrawn from Themelis and Uloa, 2007.)

The generation of electricity also appears to be the main use in the USA (Fig. 5.3). In the USA, 3.7 billion Nm^3 of methane is produced in landfill sites generating 1071 MW of power (Themelis and Ulloa, 2007).

Anaerobic Digestion

The anaerobic breakdown of organic material has only been studied in detail in the case of the degradation of sewage sludge and an outline of the process is given in Fig. 5.4. A consortium of microorganisms that develop under anaerobic conditions degrade the organic materials in series of stages.

In the first stage is the hydrolysis phase. Here complex organic materials consisting of carbohydrates, lipids, proteins, DNA and RNA are broken down by hydrolytic bacteria such as *Clostridium* sp., *Eubacteria* sp. and bacteroids. The result of the hydrolysis is simple sugars, acids, ketones and amino acids. In the next stage, acidogenesis, the simpler compounds are then broken down to acetate, lactate, propionate, ethanol, carbon dioxide and hydrogen. In the next stage, acetogenesis, the simple compounds are converted into acetate. Acetate is then combined with carbon dioxide

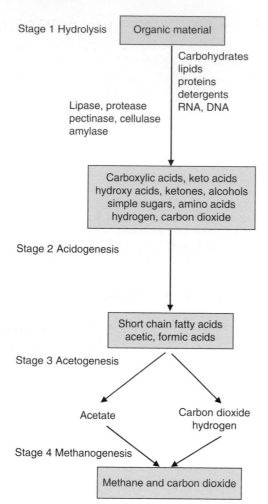

Stage 1 Hydrolysis

Organic material

Carbohydrates
lipids
proteins
detergents
RNA, DNA

Lipase, protease
pectinase, cellulase
amylase

Carboxylic acids, keto acids
hydroxy acids, ketones, alcohols
simple sugars, amino acids
hydrogen, carbon dioxide

Stage 2 Acidogenesis

Short chain fatty acids
acetic, formic acids

Stage 3 Acetogenesis

Acetate

Carbon dioxide
hydrogen

Stage 4 Methanogenesis

Methane and carbon dioxide

Fig. 5.4. Stages of the anaerobic breakdown of organic materials.

and hydrogen to form methane by a group of bacteria known as the methanogens in the last stage, methanogenesis. These bacteria are some of the most oxygen-sensitive bacteria found and include *Methanobacterium* sp., *Methanobacillus* sp., *Methanococcoides* sp. and *Methanosarcina* sp. The methanogens function in close contact with the acetogenic bacteria in time and space. This allows any hydrogen formed to be transferred without loss to the atmosphere. Strict anaerobes only grow and metabolize slowly so that anaerobic metabolism is much slower than aerobic metabolism and this is why the process can take up to 30 days. There are other processes that can lead to the production of methane, carbon dioxide and hydrogen.

Methanococcoides and Methanolobus form methane from acetate which can be the preferred substrate in cold wet anaerobic soils found in wetlands:

$$CH_3COOH \rightarrow CH_4 + CO_2 \tag{5.1}$$

Another reaction which can occur converts carbon monoxide produced from acetate to carbon dioxide and hydrogen:

$$CH_3COOH \rightarrow CH_3 + CO \tag{5.2}$$

$$CO + H_2O \rightarrow CO_2 + H_2 \tag{5.3}$$

Other methanogens produce methane by a complex series of reactions involving hydrogen and carbon dioxide:

$$4H_2 + CO_2 \rightarrow CH_4 + 2H_2O \tag{5.4}$$

The exact proportion of gases varies depending on the materials broken down and the process conditions. It is likely that similar stages occur in landfill and anaerobic digestion of animal slurries:

Sewage Sludge

Sewage sludge consists principally of microorganisms and organic materials and is a by-product of the aerobic treatment of sewage. Sewage sludge can be disposed of on land, dried, put in landfill, composted, or incinerated, but one of the most efficient methods of disposal is to anaerobically digest the sewage sludge. The advantage of anaerobic digestion is that it reduces the volume of sludge, produces biogas and as it is enclosed less smell is associated with the process. There are a number of anaerobic digester designs but the simplest design is a large, sealed vessel of 200,000–400,000l which is maintained at 30–37°C for 30 to 60 days (Fig. 5.5). The sewage sludge is broken down by anaerobic microorganisms producing a gas which is predominantly methane (biogas) and also contains carbon dioxide and hydrogen. A digester can produce gas at a rate of

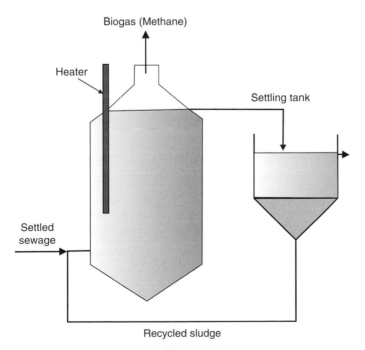

Fig. 5.5. Anaerobic treatment of sewage sludge in simple sealed vessel. (From Scragg, 2005.)

$1.0\,m^3/m^3/day$ that represents between 4.2 and 10.4 GJ. The biogas collected has a reduced calorific value compared with 100% methane as it contains carbon dioxide.

Biogas from Landfill Sites

Landfill sites are used to dispose of a wide range of waste materials from domestic and industrial sources which contain organic materials. The composition of municipal solid waste (domestic) is given in Table 5.1. As the landfill is sealed, the conditions within the site soon become anaerobic, producing biogas. The stages of anaerobic degradation of the organic content are probably the same as for sewage sludge. In older landfill sites, the biogas generated was allowed to escape but at present the biogas is captured and used. In 2001 it was reported that there were 955 landfill sites globally where the gas was recovered.

The composition of municipal solid waste is very likely to change over time as recycling and reuse of waste continues to increase. In the UK, the composting of garden waste and the recycling of metals, glass and plastics along with degradable plastic will reduce the organic composition of landfill waste. The cost of landfill is also rising which will also reduce the amounts of waste which are placed in landfills.

In the construction of landfill sites an impermeable barrier to stop any leachate reaching the groundwater is required. The most suitable sites are abandoned quarries and opencast sites preferably with a non-porous substratum. The site is lined with clay, plastics and rubber and once sealed the site can be filled in cells or terraces (Fig. 5.6), each cell or terrace being covered with soil after compaction at the end of each day. The compaction reduces the amount of air trapped, helping anaerobic conditions form and avoiding spontaneous combustion by reducing oxygen content. In order to collect the biogas generated during construction, permeable horizontal trenches or perforated pipes are incorporated into the terraces and cells (Fig. 5.7). Once a landfill has reached its working level it is capped with clay, a drainage layer and soil. The drainage layer stops rainwater entering the landfill, reducing the leachate formed. Once capped the organic material in the landfill will begin to degrade and although aerobic at the start, oxygen will be soon exhausted and conditions will become anaerobic.

Table 5.1. Composition of municipal solid waste as percentages. (From Scragg, 2005; Themelis and Ulloa, 2007.)

Composition (%)	UK waste	US waste
Organic waste		
Paper	35–60	32–36.2
Garden waste	2–35	12.1–14.8
Food waste	2–8	8.5–11.7
Wood	1–3	5.8
Textiles	1–3	3.7
Inorganic		
Metals	6–9	6.3
Glass	5–13	6.4
Plastics	1–2	11.8

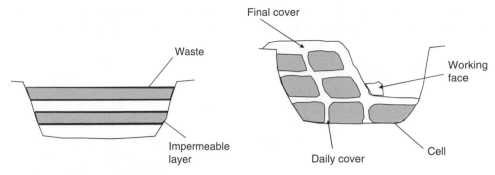

Fig. 5.6. Construction of landfill sites, where each cell or terrace is covered with soil after compaction. (From Scragg, 2005.)

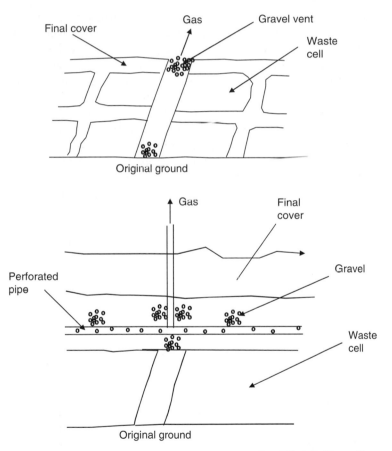

Fig. 5.7. The construction of the gas collection system in a landfill site. (From Scragg, 2005.)

The biogas collected will, like that produced from sewage sludge, have about half of the calorific vale of natural gas because of the presence of carbon dioxide and the composition will vary depending on the waste composition in the landfill. The gas is normally extracted 1–2 years after capping of the site and at best yields $100\,\text{m}^3$ gas per tonne of waste. The total extracted is only 25% of the possible yield because of the slow rate of gas formation and migration within the site. Figure 5.8 shows the changes in the landfill gases as the site develops and the organic material is degraded. A typical timescale of gas production from a landfill site is given in Fig. 5.9.

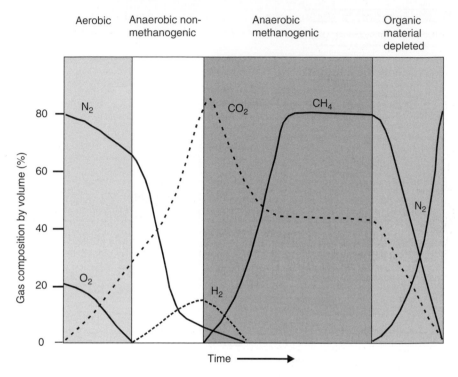

Fig. 5.8. Changes in landfill gases over time as the organic material is degraded.

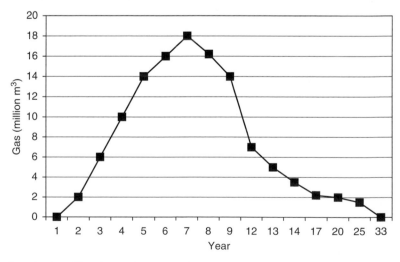

Fig. 5.9. A typical profile of gas production in millions of cubic metres from a landfill site. (Redrawn from NSCA, 2006.)

Biogas from Anaerobic Digestion of Agricultural Wastes

Small anaerobic digesters have been installed on farms to treat excess animal slurries which cannot be placed on the land. The biogas formed is normally used for heating but can also be used in dual-fuel engines.

In a number of developing countries, such as China and India, both animal and human wastes are anaerobically digested on site. The biogas formed is used as a low pressure source of gas for domestic use.

Use of Biogas as a Transport Fuel

The gas produced by anaerobic digestion of wastes consists mainly of methane (50–75%) and smaller amounts of carbon dioxide and hydrogen, and has an energy content of 20–25 MJ/kg, less than that of 100% methane which has energy content of 50.2 MJ/kg (Table 5.2). However, this is sufficient energy to be used in boilers and engines but if it is to be used as a transport fuel it will need to be upgraded to ~95% methane. The upgrading consists of the removal of contaminants such as carbon dioxide, hydrogen sulfide, ammonia, particles and water so that the gas contains 95–98% methane. The most common methods available to remove carbon dioxide are water scrubbing and pressure swing absorption. Hydrogen sulfide can be removed during anaerobic digestion by adding iron chloride and air-oxygen dosing of the resulting gas.

Light-duty vehicles running on biogas will normally be fitted to petrol engines where the vehicle retains the ability to run on petrol. In contrast, heavy-duty vehicles are normally run on biogas only. The gas is stored compressed or liquefied where compressed gas is the most common at 200 bar. The energy density of compressed gas is much less than in liquids so that the vehicle range is reduced or the tank needs to be much larger (Table 5.2). In the heavy-duty vehicles where long range is important, liquefied gas is generally used. As a transport fuel methane, like LPG, CNG and LNG, has seen only limited use because of the costs of modification and installation, despite being a cheaper fuel due to tax concessions. The number of alternative fuelled vehicles in the USA from 1993 is shown in Fig. 5.10. The number of CNG vehicles has remained static as has LNG vehicles, whereas the number of LPG vehicles has declined. There has been a slow increase in electric vehicles where the number excludes electric hybrids, and a very small number of hydrogen power vehicles. It is those vehicles capable of running on ethanol E85 (85% ethanol) that have shown a rapid increase and the EIA estimates that there are some 6 million vehicles capable of using E85.

The Energy Saving Trust (2007) states that in 2007 there were 1490 LPG stations and 18 dispensing natural gas (NG) in the UK but by comparison there were only 622 in Germany and 521 in Italy.

The properties of the transport fuel gases methane, propane and butane are compared with diesel and petrol in Table 5.2. Methane has a higher energy content than petrol but a much lower octane number. The octane number is a measure of the resistance of the fuel to pre-ignite when compressed. A low octane fuel will pre-ignite causing a condition known as 'pinking' with a loss of power.

Table 5.2. Properties of methane, propane, butane, petrol and diesel.

Properties	Methane (CH_4)	Hydrogen (H_2)	Propane (C_3H_8)	Butane (C_4H_{10})	LPG 60% propane 40% butane	LNG	Petrol (C_7H_{16})
Molecular weight	16.07	2	44.11	58.13	–	–	100.2
Carbon (%)	37.5	0	82	96	–	–	85–88
Liquid density (kg/l)	0.72	0.071	0.5	0.58	0.5	0.5	0.74
Gas density (kg/l)	0.0007	0.000084	0.0018	0.002	0.002	0.5	–
Liquid energy (MJ/kg)	50.2	141.9	46.4	45.7	48.8	50.0	44.0
Gas energy (MJ/l)	0.039	0.012	0.079	–	0.024	0.023	–
Boiling point (°C)	–161	–253	–42	–0.5	–	–	35–200
Cetane number	5	–	5	20	–	–	0.5
Octane number	10	–	112	–	–	–	90–100
Sulfur (%)	0	0	0	0	0	0	0.05

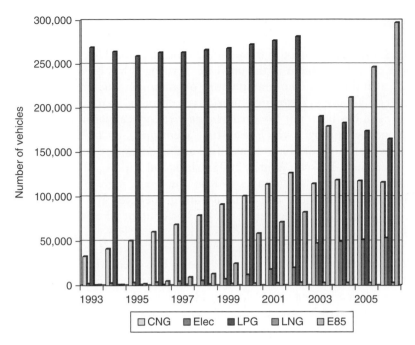

Fig. 5.10. Estimated number of alternative-fuelled vehicles in the USA. (Redrawn from EIA, 2008.)

Whether biogas is to be used for electricity generation or as a transport fuel, the total biogas available in the UK at present has been estimated as just over 6 Mtoe (Table 5.3). In 2006, the UK used over 60 Mtoe as transport fuels, 37.2 Mtoe in electricity and 232.2 Mtoe in total. So biogas can make a contribution to the renewable portion of energy used in the UK but the contribution will only be small at 2.7%.

Table 5.3. Total methane (biogas) potential in the UK produced by anaerobic digestion. (Adapted from NSCA, 2006.)

Material	Tonnes/year dry	Gas factor m³/tonne	Total methane	Tonnes of oil equivalent
Sewage sludge	1,400,000	195	273,000,000	231,400
Wet farm slurries				
Dairy	2,016,000	130	262,080,000	222,144
Pig	535,000	195	104,325,000	88,428
Poultry	1,515,000	236	357,918,750	303,379
Dry manure				
Cattle	6,253,140	160	1,000,502,400	848,045
Pig	4,532,414	180	815,834,520	691,517
Horses	458,172	75	34,362,900	29,127
Commercial food waste	6,295,000	330	2,077,350,000	1,760,801
Domestic food waste	7,510,644	330	2,478,512,520	2,100,834
Total	30,515,370		7,403,886,090	6,275,675

Hydrogen

Hydrogen is the first element in the periodic table, a colourless, odourless gas which is the most plentiful element in the universe. Hydrogen has been used extensively in the chemical industry in the manufacture of ammonia, methanol, petrol, heating oil, fertilizers, vitamins, cosmetics, lubricants, cleaners, margarine and as a rocket fuel. Hydrogen has been put forward as a new energy carrier in a system known as the 'hydrogen economy', which was first mentioned in 1972. In the hydrogen economy, hydrogen would be used as a fuel and to transport and store energy in the way that electricity is used (Winter, 2005; Clark and Rifkin, 2006). Hydrogen as an energy source has many advantages as it is non-toxic, high in energy, on combustion yields only water and it can be used both in fuel cells and internal combustion engines. Hydrogen has three times the energy content of petrol and methane at 141.9 MJ/kg but because of its low density it has very low energy content per unit volume (Table 5.4). Table 5.4 compares the energy per unit mass and unit volume for hydrogen, petrol and other gaseous fuels. It is clear that both methane and hydrogen in the gaseous state have low energy per unit volume but in hydrogen in liquid form the energy per unit volume was still low. However, there are disadvantages in the production, storage and flammability of hydrogen which have been used to question the adoption of hydrogen as an energy carrier (Hammerschlag and Mazza, 2005).

There are a number of chemical and biological routes for the production of hydrogen, but only some processes are renewable and sustainable. These methods are listed below where only the first three methods are operated at an industrial scale.

Non-renewable:

- Steam reformation of methane (natural gas).
- Coal gasification.
- Partial oxidation of heavy oil.
- Thermocatalytic treatment of water.

Renewable:

- Electrolysis of water using electricity, only renewable if sustainable electricity supply used such as wind or solar power.
- Photocatalytic splitting of water using TiO_2.

Table 5.4. Comparison of the energy content of liquid and gaseous fuels. (Adapted from Midilli *et al.*, 2005.)

Fuel	Energy content mass (MJ/kg)	Energy content volume (MJ/l)
Petrol (liquid)	47.4	34.8
LPG (liquid)	48.8	24.4
LNG (liquid)	50.0	23.0
Hydrogen (liquid)	141.9	11.9
Hydrogen (gas)	141.9	0.012
Methane (liquid)	50.2	36.4
Methane (gas)	50.2	0.039

- Gasification of food waste, sewage sludge, biomass.
- Pyrolysis of biomass.
- Biological processes:
    1. anaerobic metabolism (dark fermentation);
    2. photosynthetic hydrogen production (direct biophotolysis);
    3. indirect hydrogen production;
    4. photo-fermentation;
    5. carbon monoxide metabolism, water–gas-shift reaction.

The most widely used process for producing hydrogen is steam reforming of natural gas which is the least expensive process and produces 97% of the hydrogen made. The other two commercial processes are the gasification of coal (Chapter 4) and the partial oxidation of heavy oils. Clearly none of these processes are renewable and sustainable.

There are a number of renewable methods of producing hydrogen but to date these are all at the experimental stage (Kapdan and Kargi, 2006). The theme of the book is biofuels and the production of energy from biological materials and therefore this section will concentrate on the biological production of hydrogen. As can be seen from the list there are a number of biological processes which result in hydrogen.

Production of hydrogen from biological material

Gasification of food waste, sewage sludge and biomass

If biomass is gasified at temperatures above 700°C, a mixture of gases and charcoal is produced. The gas produced contains mainly hydrogen, carbon monoxide, methane, carbon dioxide and nitrogen. More hydrogen can be produced by a water-shift reaction where carbon monoxide is reacted with water to form carbon dioxide and hydrogen:

$$\text{Biomass} + \text{heat} + \text{steam} \rightarrow H_2 + CO + CO_2 + CH_4 \quad (5.5)$$
$$+ \text{ hydrocarbons} + \text{charcoal}$$

Water-shift reaction

$$CO + H_2O = CO_2 + H_2 \quad (5.6)$$

Using a fluidized bed gasifier with a catalyst it has been possible to obtain 60% hydrogen production. The main problem with gasification is the formation of tar, which can be minimized by gasifier design, control and additives (Ni *et al.*, 2006). The hydrogen can be separated from the other gases by pressure swing adsorption.

Pyrolysis

Pyrolysis involves the heating of biomass at 300–500°C at 0.1–0.5 MPa in the absence of air. This produces liquid tars and oils, charcoal and gases including hydrogen, methane, carbon monoxide and carbon dioxide. The water-soluble pyrolysis oil can be used for hydrogen production by catalytic steam reforming (water shift). Methane can be steam reformed to produce more hydrogen:

$$CH_4 + H_2O = CO + 3H_2 \tag{5.7}$$

The carbon monoxide can also be used to form hydrogen in a water-shift reaction:

$$CO + H_2O = CO_2 + H_2 \tag{5.8}$$

Biological production of hydrogen

The biological production of hydrogen has been known since the early 1900s and the enzymes involved were discovered in the 1930s. Hydrogen production has been found in many prokaryotes, green microalgae, and a few eukaryotes as shown in Table 5.5 (Das and Veziroglu, 2001; Happe *et al.*, 2002).

The production of hydrogen is due to the presence of two enzymes either nitrogenase or hydrogenase in the organism. Nitrogenase has the ability to use ATP and electrons to reduce substrates including protons to hydrogen gas and has been found in photoheterotrophic bacteria such as *Rhodobacter* sp.:

$$2e^- + 2H^+ + 4ATP \overset{\text{nitrogenase}}{=} H_2 + 4ADP + 4Pi \tag{5.9}$$

Hydrogenases have been found in a large number of green microalgae such as *Chlamydomonas reinhardtii* and *Chlorococcum littorale*, anaerobic bacteria such as *Clostridium* sp. and *Cyanobacteria* sp. Hydrogenases can be either uptake or reversible hydrogenases and can be divided into three classes based on their metal composition. These classes are Ni/Fe, Fe and metal-free where the Fe hydrogenase has a unique active centre giving the enzyme a 100-fold higher activity:

Table 5.5. Microorganisms capable of producing hydrogen. (Adapted from Das and Veziroglu, 2001.)

Microbial type	Example	Comments
Green algae	*Scenedesmus obliquus* *Chlamydomonas reinhardtii*	Produce hydrogen from water using solar energy. Inhibited by O_2
Cyanobacteria (heterocysts)	*Anabena azollae* *Nostoc muscorum*	Nitrogenase enzyme produces hydrogen, the enzyme in heterocyst protected from O_2 inhibition. Requires light. Fixes nitrogen
Cyanobacteria (non-heterocysts)	*Plectonema boryanum* *Oscillotoria limnetica*	As above, but no protection of nitrogenase
Photosynthetic bacteria	*Rhodobacter sphaeroides* *Chlorobium limicola*	Can use waste, require light
Anaerobic bacteria (fermentation)	*Clostridium butylicum* *Desulfovibrio vulgaris*	Functions in the dark, can use a variety of substrates

$$H_2 \overset{\text{hydrogenase}}{\leftrightarrow} 2H^+ + 2e \qquad\qquad (5.10)$$

These hydrogenases and nitrogenase are responsible for hydrogen production by a number of microorganisms.

Anaerobic metabolism (dark fermentation)

Anaerobic breakdown of organic material can yield hydrogen in a number of cases. Anaerobic digestion of sewage sludge by a consortium of microorganisms can produce small amounts of hydrogen in addition to the major product biogas (section 'Anaerobic digestion', Chapter 5, this volume). Anaerobic digestion or fermentation of organic compounds by *Clostridium* sp. and some microalgae was also found to produce hydrogen under specific conditions. One of the best-known examples is the production of acetone and butanol by *Clostridium acetobutylicum* growing anaerobically on glucose (molasses). This process was used from 1915 until the 1950s to produce acetone and butanol for the munitions and chemical industries. The biological process has now been replaced by the production of acetone and butanol from petrochemicals. However, other products are formed along with acetone and butanol and include ethanol, butyrate, acetic acid, carbon dioxide and hydrogen. Depending on the culture conditions, and the strain used, the amount of the various products formed can vary including the amount of hydrogen produced. The pathway involved in the production of acetone and butanol is shown in Fig. 5.11. It has been estimated that 2 mol of hydrogen are formed per mole of glucose consumed (Ni *et al.*, 2006).

Green algae such as *C. reinhardtii* respond to anaerobic conditions or nutrient reduction (sulfur) by producing hydrogen. The induction of anaerobic conditions switches the organism's metabolism to fermentative which produces a number of harmful end products such as ethanol and organic acids. Under these conditions hydrogenase activity is inhibited and hydrogen acts as an electron sink, avoiding some of the problems of aerobic conditions. The low sulfur condition causes the downregulation of the photosystem II where the lack of sulfur-containing amino acids blocks the repair cycle for photosystem II.

Photosynthetic hydrogen production (direct biophotolysis)

In photosynthesis, solar energy is used by photosystem II to split water and release oxygen, electrons and protons. Photosystem I uses solar energy to produce the reducing power required to fix carbon dioxide, and the electrons are passed along the electron transfer system, eventually generating ATP. In the direct use of solar energy to convert water into hydrogen the electrons are transferred along the electron transfer chai until the penultimate step catalysed by ferredoxin, where the electrons are transferred to a hydrogenase, converting protons into hydrogen (Fig. 5.12). These reactions are carried out by green microalgae and blue-green *Cyanobacteria* sp.

However, hydrogenases are inhibited by oxygen so that the concentration of oxygen needs to be kept below 0.1%. In the case of *C. reinhardtii* oxygen is removed

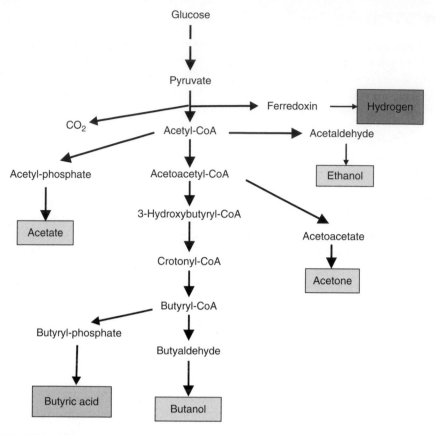

Fig. 5.11. The pathway involved in the production of acetone and butanol by *Clostridium acetobutylicum.*

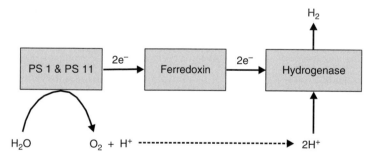

Fig. 5.12. The direct production of hydrogen using light carried out by green algae and Cyanobacteria.

by respiration but because substrate is consumed the efficiency is low. In some cyanobacteria such as *Anabena cylindrica* photosynthesis is split between two types of cells. Photosystem II functions in the vegetative cells whereas the heterocysts contain the carbon-fixing portion and a hydrogenase enzyme which is protected from the inhibitory effects of oxygen by a thick cell wall.

Indirect hydrogen production

In this case, growing in the light photosynthesis is used for growth and to store carbohydrates. When the organism is switched to aerobic dark conditions the stored carbohydrates or cell material is metabolized in the same way as in Fig. 5.12, yielding hydrogen. This type of two-stage process has been observed in *Cyanobacteria* sp.

Photo-fermentation

Some photoheterotrophic bacteria (purple non-sulfur bacteria) such as *Rhodobacter* sp. and *Rhodospirillum* sp. convert organic acids in the light into carbon dioxide and hydrogen (Fig. 5.13). The key enzyme in these organisms is nitrogenase which requires ATP to produce hydrogen. The nitrogenase is inhibited by oxygen, ammonia and high nitrogen to carbon ratios so that oxygen-free conditions are required.

Carbon monoxide metabolism (water-shift reaction)

Some photoheterotrophic bacteria, for example *Rhodospirillum rubrum*, can metabolize carbon monoxide in the dark in a reaction similar to the water-shift reaction:

$$CO + H_2O = CO_2 + H_2 \tag{5.11}$$

Hydrogen Storage

One of the main problems with the use of hydrogen as a transport fuel and energy carrier is how to store it as it is a light gas (Coontz and Hanson, 2004; Zhou, 2005). Compression and subsequent storage at high pressure in cylinders is a common method of storing gases. However, the density of hydrogen is low ($0.089 \, kg/m^3$) compared with methane ($0.717 \, kg/m^3$) and therefore a pressure some four times higher than normal is required (345–690 bar; 10,000 psi) to contain sufficient hydrogen.

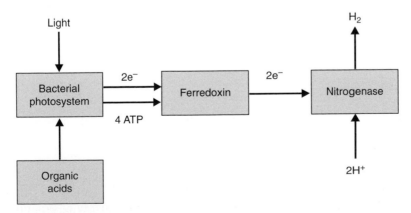

Fig. 5.13. Photo-fermentation production of hydrogen.

Even at these pressures it still requires the fitting of a tank eight times the size of the equivalent petrol tank, and tanks of this size are not available commercially. Therefore, compressed gas is not likely to be used at present.

Another common method of storing gas is as a liquid. Hydrogen has a boiling point of −253°C and the critical temperature for liquid hydrogen is −240°C. Therefore, the liquid hydrogen will need to be stored in well-insulated tanks in an open system to avoid pressure build-up and thus some gas will be lost on storage. Whether compression or liquefaction is used both require the input of energy, between 214 and 354 MJ/m^3 for compression and 15.2 kWh/kg for liquefaction which is some 30% of the energy contained in the fuel (Midilli *et al.*, 2005; de Wit and Faaij, 2007).

Despite these problems the USA has invested US$1.7 billion and the €2.8 billion in hydrogen power vehicles. Ford, Mazda and BMW have produced hydrogen-powered internal combustion engines and Honda, Ford, Toyota, General Motors, Daimler-Chrysler, Renault-Nissan and Volkswagen have produced fuel cell vehicles. Liquid hydrogen-filling stations have been installed in Munich and London. In 2006 there were 159 hydrogen vehicles in the USA not including electric hybrid vehicles (Fig. 5.10).

Because of the need to store hydrogen inexpensively, alternatives to compression and liquefaction have been investigated including complex hydrides, metallic hydrides, physisorption and nanostructures (Zhou, 2005).

Complex hydrides form between hydrogen and group I, II and III elements such as lithium, magnesium, boron and aluminium. The complex hydrides have a high hydrogen density (150 kg/m^3) and can release hydrogen at moderate temperatures. This method is still under development as conditions for hydrogen release changes the particle and the effects of repeated adsorption/desorption need to tested.

Some metals absorb hydrogen, forming hydrides. These are usually a rare-earth metal such as lanthium combined with a transition metal such as nickel. Hydrogen density has reached 115 kg/m^3 for the metal hydride $LaNi_5H_6$. In simple hydrides the metal can absorb and release the hydrogen at room temperature but their hydrogen density is low. The formation of hydrides is an exothermic reaction and the more stable the hydride the more heat is required to release the hydrogen.

Gases can be adsorbed on to a number of adsorbents in a variety of ways provided the gas is below its critical point. The gas can be adsorbed as a single layer on the surface which depends on the surface area of the material and the temperature. Adsorption decreases with a rise in temperature.

Hydrogen can be stored in carbon nanotubes but the capacity appears much lower than was first estimated. As adsorption is dependent on surface area, it is difficult for nanotubes to compete with super-activated carbon.

Hydrogen Use

Hydrogen can be used to store energy, as a fuel for the internal combustion engine, gas turbine and the fuel cell (see section 'Fuel cells', Chapter 3).

Energy storage

There are a few examples of hydrogen being used to store energy in situations where excess electricity is being generated from sustainable sources such as wind and solar power. These energy sources are intermittent and to balance out the dips in electricity production the stored hydrogen is used to fuel a generation system. One example is the island of Utsira off Norway where electricity is provided by wind power. Surplus electricity is used to electrolyse water and the hydrogen formed stored compressed in a large tank. When the wind does not blow the stored hydrogen is used in a modified internal combustion engine to drive a generator.

Another example was a combined photovoltaic and wind electricity generation system in Cooma, Australia (Shakya *et al.*, 2005). Here again the surplus electricity was used to electrolyse water and the hydrogen stored compressed in a cylinder with a capacity of $5.5\,m^3$ at 24.5 MPa (3552 psi). It was found that the electricity generated was more expensive than grid-connected electricity, as expected, but costs could be reduced as over 50% of the capital costs were the electrolyser and purification components. A reduction in the cost of these components would reduce costs considerably. The system still has great merit in situations where no grid supply is available.

Hydrogen as a transport fuel

The advantage of hydrogen as a fuel in the internal combustion and gas turbine engine is the product of combustion is only water and the exhaust contains no carbon dioxide, sulfur dioxide or carbon monoxide. Burning hydrogen does however produce NO_x which is a function of the temperature where hydrogen burns at a little higher temperature than petrol. This is not a problem encountered with fuel cells. Research is under way using recirculation of exhaust gas to reduce NO_x emissions (Heffel, 2003). Both liquid hydrogen and fuel cell vehicles have been developed by a number of automotive companies. The problems with hydrogen are its supply to vehicles and storage on board.

If we consider that hydrogen may be used as a universal transport fuel either in an internal combustion engine or in a fuel cell, then a new infrastructure will be required. The main question is what needs to be developed first, the hydrogen power vehicles or the infrastructure? The widespread use of hydrogen-powered vehicles will need a network of filling stations but these will probably not be built until there is a number of vehicles to justify the cost of construction. The most likely sequence will be the construction of the infrastructure prior to the widespread introduction of hydrogen-powered vehicles. A study of the early development of the fossil fuel automotive industry illustrates the same problems (Melaina, 2007). The car was introduced via mass production but in the beginning there were few petrol stations, as is the case for hydrogen at present. However, motorists could obtain fuel in cans from shops and repair garages. This supply system had developed for the sales of paraffin (kerosene) for lighting and heating before the introduction of cars. As the car ownership increased there was a demand for filling stations which were built in increasing numbers. Such is the nature of compressed or liquid nitrogen that small-scale supplies cannot be obtained without the development of a widespread infrastructure.

Some of the options for the supply of hydrogen to filling stations are shown in Fig. 5.14. The first option is to produce hydrogen, in this case from natural gas, at a central production unit. The hydrogen is liquefied and transported to the filling station by tanker where it is stored until required. This is very much the system that is

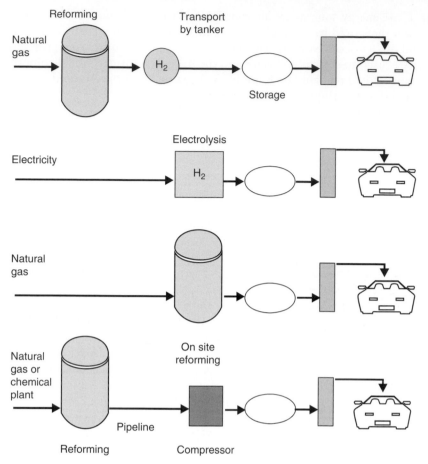

Fig. 5.14. Some of the possible methods of supplying hydrogen to petrol stations.

used at present for petrol and diesel. In the second option the hydrogen is generated at the filling station either by electrolysis or from natural gas and stored. In the third option hydrogen is supplied to the filling station by pipeline where it is compressed or liquefied. All these options have some of the problems outlined above, including the energy required to compress or liquefy the hydrogen, its storage and its dispension. A well-to-wheel assessment of the supply of hydrogen and storage on the vehicle concluded that the energy used to deliver hydrogen in a liquid form was similar to that for petrol and diesel.

The on-site generation was much more costly in terms of energy. In both cases 80% of the energy use was expended on hydrogen production and liquefaction or compression (de Wit and Faaij, 2007). When the fuel delivery and driving costs were considered, on-site generation was again the most expensive. The driving costs were higher than petrol and diesel in all cases.

Once the hydrogen has been made available there are a number of options for its use in the vehicle for both internal combustion engines and fuel cells (Ogden *et al.*, 1999). Compressed hydrogen or liquid hydrogen will have to be stored in

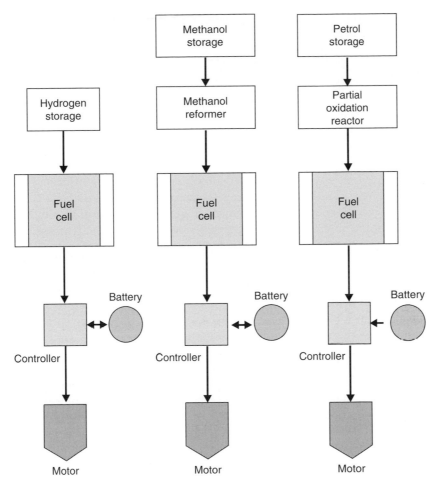

Fig. 5.15. The options available for the supply of hydrogen to fuel cells in cars.

cylinder/tanks at pressure of up to 600 psi or at –253°C in insulated tanks. Either method of storage will pose problems when filling the vehicle. There are options to avoid the on-board storage of hydrogen which includes the on-board production of hydrogen from methanol or petrol (Fig. 5.15). Both methanol and petrol are liquid and can easily be supplied by the present infrastructure, but petrol is non-sustainable and methanol would need to be produced in a sustainable manner. Both systems would add weight, complication, increase fuel consumption and cost to the vehicle.

Dimethyl Ether (DME)

Dimethyl ether is a simple ether formula CH_3OCH_3 which has properties similar to propane, butane and LPG (Table 5.6). Dimethyl ether is volatile, non-toxic, non-mutagenic, non-carcinogenic, has a sweet ether odour and has been regarded as being

Table 5.6. Properties of propane, butane, dimethyl ether and diesel. (From Semelsberger *et al.*, 2006; Cocco *et al.*, 2006.)

Properties	Propane (C_3H_8)	Butane (C_4H_{10})	Dimethyl ether (CH_3OCH_3)	Diesel ($C_{14}H_{30}$)
Molecular weight	44.1	58.13	46.07	586
Carbon (%)	82	96	52.2	86
Density (kg/l)	0.5	0.58	0.66	0.86
Energy (MJ/kg)	46.4	45.7	28.6	38.5–45.8
Boiling point (°C)	−42	−0.5	−24.9	125–400
Cetane number	5	20	55–60	40–55
Sulfur (%)	0	0	0	0.2

environmentally benign (Semelsberger *et al.*, 2006). As DME is non-toxic and non-corrosive it is used mainly as a hairspray propellant, in cosmetics and in agricultural chemicals. DME has a high cetane value, no sulfur, little particulate matter (PM) emissions and can be competitive with LPG. Owing to these characteristics DME has been considered as a fuel for diesel engines, gas turbines and fuel cells. As its properties are similar to those of propane and butane (Table 5.6), DME could be used to replace or supplement LPG for distributed power generation including gas turbines (Cocco *et al.*, 2006). DME can also be reformed to produce hydrogen for fuel cells.

Dimethyl ether is produced in a two-step process where methanol is produced from syngas which is normally produced by the steam reformation of methane (natural gas). The methanol is then dehydrated to form dimethyl ether. Syngas can be produced from waste and biomass so that DME could be produced from sustainable sources. To produce methanol the syngas needs a 1:1 ratio of carbon monoxide to hydrogen which can be adjusted by the water-shift reaction:

$$CO + 2H_2 \leftrightarrow CH_3OH \tag{5.12}$$

$$2CH_3OH \leftrightarrow CH_3OCH_3 + H_2O \tag{5.13}$$

DME has been shown to produce low noise, smoke-free combustion, and reduced NO_x when used in an internal combustion engine (Huang *et al.*, 2006). Because of its high cetane number and low boiling point, DME has been used at 100% or as an oxygenated addition to diesel. However, DME requires special fuel handling and storage as its properties are similar to LPG, and the lower energy means that a larger fuel tank will be required. The engine does not require modification but as the viscosity of DME is so low it can cause leakage in the pumps and injectors. Another consequence of the low fuel viscosity is a reduction in lubrication where lubricants need to be added to the fuel if used for long periods. The conclusions on DME were that it gave lower NO_x and SO_x, is soot-less and produces the highest well-to-wheel value compared with FT-biodiesel, biodiesel, methanol and ethanol (Semelsberger *et al.*, 2006).

DME has also been shown to be suitable for gas turbines where performance and carbon dioxide emissions were improved when it was used in a chemically recuperated gas turbine (CRGT) (Cocco *et al.*, 2006). One of the options to increase the efficiency of gas turbines is to recover the exhaust heat chemically. Most CRGT systems use methane steam reforming where carbon dioxide and hydrogen are formed and used as a fuel in the turbine. However, a high reforming temperature up to 600–800°C is required, which is higher than the exhaust temperature of commercial gas turbines. On the other hand methanol, DME and ethanol have lower reforming temperatures of 250–300°C, 300–350°C and 400–500°C respectively which makes them more suitable. The overall reforming process is described below and a CRGT system is shown in Fig. 5.16:

$$CH_3OCH_3 + H_2O \leftrightarrow 2CH_3OH \qquad (5.14)$$

$$CH_3OH \leftrightarrow CO + 2H_2 \qquad (5.15)$$

$$CO + H_2O \leftrightarrow CO_2 + H_2 \qquad (5.16)$$

The reformation is carried out at 290°C using Cu/SiO_2 and HPA/Al_2O_3 catalysts. The CRGT system achieves an efficiency of 54%, 44% higher than a standard plant.

DME can also be used as a fuel for fuel cells as it can be reformed to produce hydrogen. The reformation is in two steps where the first step is an acid catalysis which converts dimethyl ether into methanol and the second step is the reformation of methanol over a Cu or Cu/Zn catalyst:

$$CH_3OCH_3 + H_2O \leftrightarrow 2CH_3OH \qquad (5.17)$$

$$2CH_3OH + 2H_2O \leftrightarrow 6H_2 + 2CO_2 \qquad (5.18)$$

Other similar compounds have been tested in diesel engines which include dimethyl carbonate and dimethoxy methane (Huang *et al.*, 2006).

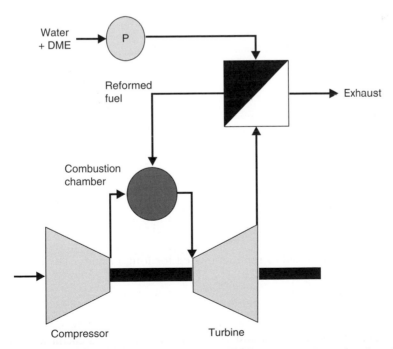

Fig. 5.16. A chemically recuperated gas turbine (CRGT) system using reforming dimethyl ether (DME) to utilize the exhaust heat. (From Cocco *et al.*, 2006.)

Conclusions

Gaseous biofuels are a mixture of existing technologies and the promise of technology in the future. Methane is produced in landfill and anaerobic digesters and used for heating and electricity generation. Methane as a transport fuel has been proposed and the technology for the compression and liquefaction of the gas exists but may not be implemented. Methane is similar to LPG. LPG has been around for some years but the take-up of cars using this fuel has been very slow, probably due to the required expensive modifications to the car and the uneven distribution of LPG filling stations.

Much has been written about hydrogen and the 'hydrogen economy' but considerable advances in technology will be required to make hydrogen a transport fuel that is sustainable. At present all hydrogen is made from natural gas and to be sustainable it needs to be produced from renewable resources such as biomass. The problem with hydrogen as a fuel is it needs to be stored so that sufficient fuel can be carried in a vehicle. At present, hydrogen can either be compressed gas or liquefied but both require considerable amounts of energy and special tanks. The production and distribution of hydrogen will also be required if it is to be used as a transport fuel. This will need the installation of a completely new infrastructure at a considerable cost. Despite the problems there is probably a future for hydrogen as a fuel for fuel-cell-powered vehicles; the question is whether hydrogen will be produced on-board or stored as hydrogen.

6   Liquid Biofuels to Replace Petrol

Introduction

Oil makes up 35% of the world's primary energy supply and the majority of this oil is used to produce the transport fuels petrol, diesel and kerosene. If biofuels are to be used to replace the liquid fuel produced from oil, the scale of the replacement needs to be appreciated. The world used in 2005 about 1900 Mtoe oil for transport, the USA 601 Mtoe, the EU 25 342 Mtoe and the UK 58 Mtoe. The fuels used for transport in the UK, the EU 25 and world are given in Table 6.1. At present almost all transport fuels are liquid and used in internal combustion, compression ignition and jet engines. The liquid fuels are predominantly petrol, diesel and kerosene. The pattern of fuel use will vary depending on the country as many use more diesel than petrol. In the EU, diesel use exceeds petrol use and the more recent figures for the UK indicate that diesel use has exceeded petrol for the first time.

The possible biofuel replacements for petrol and diesel were sketched out in Chapter 4, section 'The Nature of Biofuels: First-, Second- and Third-generation Biofuels', where the three generations of biofuels were explained. This chapter deals with biofuels which can supplement or replace petrol and include methanol, bioethanol, biobutanol and FT-petrol.

Table 6.1 gives the amount of liquid fuels used for transport in the UK in 2006 and in the EU 25 and world in 2005. In the UK, diesel use has increased rapidly and now exceeds petrol, following the trend set in the EU, whereas petrol use is greater on a global basis.

Methanol

Methanol (CH_3OH) is a simple alcohol commonly known as 'wood alcohol'. It is a toxic, colourless, tasteless liquid with a faint odour which can be used in a spark ignition engine. Its characteristics are given in Table 6.2, where it is compared with petrol. Methanol contains considerably less energy than petrol but the high octane rating gives more power and acceleration. It is less flammable than petrol but burns with a nearly invisible flame, making flame detection difficult. Methanol is toxic, corrosive and as it is miscible in water a spill can be an environmental hazard, but methanol offers important emissions improvements, reducing hydrocarbons and nitrogen oxide.

In the early days of motoring, methanol was used in internal combustion engines as a blend with petrol. It was used as a motor fuel in Germany in the Second World War because of the shortage of oil (Antoni *et al.*, 2007). Because of its low energy content (19.9 MJ/kg), since the 1970s it has only been used in special cases such as Indianapolis car racing, and even the Indy cars will switch to ethanol in 2007 (Solomon *et al.*, 2007). Another use for methanol is in the production of an anti-knock agent for use in petrol. Methanol is converted into methyl-*tert*-butyl-ether

Table 6.1. Liquid fuel use in the UK in 2006 and in the EU 25 and world in 2005 (tonnes × 1000). (From IEA, 2008b; Energy Statistics, BERR, 2007.)

Fuel	UK consumption in 2006	EU 25 consumption in 2005	World consumption in 2005
LPG (liquid petroleum gas)	288	3,428	16,207
Petrol	19,918 (33.9%)	107,752 (31.5%)	876,286 (44.3%)
Petrol (aviation)	46	142	2,252
Jet kerosene	10,765 (18.3%)	51,453 (15%)	229,026 (11.6%)
Kerosene	3,457	4,909	625
Diesel	23,989 (39.3%)	178,178 (52%)	687,935 (34.8%)
Other fuels	355	1,178	161,785
Total	58,818	342,131	1,974,116

Table 6.2. The characteristics of petrol and methanol.

Characteristics	Petrol	Methanol
Boiling point (°C)	35–200	65
Density (kg/L)	0.74	0.79
Energy (MJ/kg)	44.0	19.9
Flash point (°C)	13	65
Octane number	90–100	91

(MTBE), an anti-knocking agent, by acid catalysis with isobutene. MTBE has been added to petrol replacing the lead-based compounds used previously but now banned. However, there have been concerns about the carcinogenicity and groundwater contamination by MTBE and it is being replaced by ethyl-*tert*-butyl-ether (ETBE) which can be made from ethanol.

At present, most methanol is made from natural gas but renewable sources such as woody crops, agricultural residues, forestry waste and industrial and municipal waste can be used to produce methanol by either thermochemical conversion or gasification. The gasification of the biomass at high temperatures (above 700°C) in the presence of oxygen results in a mixture of gas, tar and charcoal due to partial oxidation. The gasification process needs biomass with moisture content of 10% or below. The gas formed is called 'syngas', and is a mixture of carbon monoxide and hydrogen which can be converted into methanol if passed over Cu/Zn/Al catalysts.

A more recently developed biological method for producing methanol is the de-esterification of the methylated carboxyl groups of galacturonic acid by a pectin methyl esterase to give methanol (Antoni *et al.*, 2007). Pectin is a major component of plant cell walls and one suitable substrate for this process is sugarbeet pulp which contains 60% pectin on a dry weight basis.

Since 2000 in the USA, no cars have been run on 100% methanol, but some 15,000 M85 (85% methanol combined with 15% petrol) vehicles are in operation, mainly in California and New York. Perhaps the most promising use for methanol is in hydrogen fuel cell vehicles where it is converted into hydrogen on board the vehicle.

Ethanol

Ethanol and ethanol–petrol blends are not new as fuels for the internal combustion engine, since these fuels were proposed in the late 1800s by early car manufacturers. Henry Ford once described ethanol as the 'fuel for the future'. During the First and Second World Wars, ethanol was mixed with petrol in order to preserve oil stocks. After the First World War, petrol dominated the fuel market although ethanol still continued to be used as an octane enhancer (anti-knock) in the 1920s but this was superseded by tetra-ethyl lead. The use of ethanol as a fuel re-emerged in the 1930s in the USA, where ethanol produced from maize was sufficiently cheap to be used in blends. It was used in concentrations of 5–17.5% to produce a blend called 'gasohol' and marketed as 'Agrol'. In the UK, gasohol was marketed by the Cleveland Oil Company under the name of 'Discol' in the 1930s, a blend which continued to be sold until the 1960s. In the USA, gasohol was dropped by 1945 due to the availability of cheaper petrol.

In 1975 Brazil introduced the 'Proalcool' Programme to produce ethanol from sugarcane as a fuel to replace petrol as a response to oil price rises from 1973. The rise in oil prices was due to an Organization of the Petroleum Exporting Countries (OPEC) oil embargo as a consequence of the Arab–Israeli War in 1973. The reasons for the development of ethanol as a fuel in Brazil were to reduce the imports of petrol as Brazil had few oil fields, to open up areas of the country for cultivation, to provide employment, to increase the industrial base, and to develop ethanol exports of plant and expertise. In addition, Brazil is one of the largest producers of sugar from sugarcane so that a good substrate for ethanol production was readily available which did not require processing. The production of ethanol was encouraged by grants and subsidies to make ethanol cheaper than petrol. By the late 1980s about 50% of the cars used 95% (E95) ethanol as a fuel. However, price rises and a sugar shortage have reduced ethanol use to about 20% of vehicles, although 40% of the total fuel used is ethanol. One unforeseen outcome of the development of a large ethanol industry in Brazil producing 16.97×10^9 l in 2006 (4.49 billion gallons) is a flourishing export market for ethanol. In 2005 Brazil exported 100 million gallons to India, USA and Europe.

The USA initiated the production of fuel ethanol in 1978 with an Energy Tax Act where gasohol was defined as a blend of petrol containing more than 10% ethanol. The Act exempted ethanol from the US$0.40/gallon tax on petrol. Apart from the tax changes, support for the ethanol industry was in the form of agricultural subsidies and tax credits awarded to blenders. The driving factors for the development of an ethanol industry were similar to those in Brazil. In the case of the USA, ethanol was produced from maize starch rather than from sugarcane. In addition, the price of chemically produced ethanol in the USA increased which made biologically produced ethanol more economic. In the 1970s chemically produced ethanol was selling at US$0.145/l but in the 1980s the increase in the feedstock increased ethanol prices to US$0.53/l, which was the same price as biologically produced ethanol. The tax exemption rose to US$0.60/gallon in the mid-1980s but was reduced in 2005 to US$0.51/gallon. An additional reason for the production of alcohol as a fuel was the low prices that the farmers were getting for their maize. At present, fuel ethanol accounts for 7% of the maize crop, boosting farm incomes by US$4.5 billion and is responsible for 200,000 jobs. In the 1980s, lead in petrol was removed and ethanol was of interest to increase the octane value. The first replacement for lead in petrol was MTBE made from methanol but concerns over its toxicity has seen a change to ETBE. After considerable debate in the USA, the Renewable Fuel Standard (RFS) was signed in 2005 which required both biodiesel and bioethanol to be blended in petrol and diesel to the value of 7.5 billion gallons (US) a year by the year 2012. At the present the replacement of petrol with ethanol is also driven by the need to reduce carbon dioxide emissions, as ethanol is a carbon-dioxide-neutral sustainable product.

Properties of Ethanol

Petrol engines will run on ethanol as the properties of ethanol are similar to petrol in many aspects (Table 6.3). Butanol has been included in the table as it is an alternative to ethanol as a petrol replacement with an energy content similar to petrol. The higher heat of vaporization of ethanol means that as the fuel is vaporized in the carburettor, the mixture is cooled to a lower temperature than for petrol. This means that more fuel enters the engine, in part compensating for the lower energy content, but the fuel inlet may need heating. The octane rating is a measure of the resistance of the fuel to pre-ignite when compressed in the cylinder of the engine. A low octane fuel will pre-ignite causing a condition known as 'pinking' and this will result in a loss of power. Ethanol has a higher octane number and higher oxygen content than petrol. The heat of combustion (or gross energy) is lower than petrol, which leads to some reduction in performance and a 15–25% increase in fuel consumption.

Ethanol has the disadvantage that it mixes with water and this type of mixture will corrode steel tanks. To avoid separation of an aqueous layer in cold weather the ethanol needs to be anhydrous as ethanol normally contains 4.5% water. At 95.6% ethanol the liquid and vapour have the same concentration, known as an 'azeotrope', so no further concentration is possible by simple distillation.

Ethanol Use in Vehicles

The concentration of ethanol used in petrol differs greatly from country to country. Hydrous ethanol which contains 4.5% water (alcool) has been used in all-ethanol vehicles in Brazil, but sales of these vehicles ceased in the 1990s to be replaced with a

Table 6.3. The characteristics of petrol, bioethanol and butanol.

Characteristics	Petrol	Ethanol	Butanol
Boiling point (°C)	35–200	78	116–119
Density (kg/L)	0.74	0.79	0.81
Energy (MJ/kg)	44.0	27.2	40.5
Latent heat of vaporization (MJ/kg)	293	855	–
Flash point (°C)	13	45	37
Octane number	90–100	99	–

Table 6.4. Ethanol petrol blends used in vehicles.

Fuel	Country	Bioethanol (%)	Modification required
Alcool (E95)	Brazil	95.5	Engine modifications needed
Gasoline (E25)	Brazil	24	None
E10 (gasohol)	USA	10	None
E85	USA	85	Duel fuel cars
Oxygenate for petrol replacing MTBE	USA	7.6	None
Used in reformulated petrol	USA	5.7	None
Addition to petrol	UK	5	None

MTBE, methyl-tert-butylether.

blend containing 24% ethanol (Table 6.4). This change was probably introduced in order to avoid the modification of car engines to use 95.5% (E95) ethanol allowing the unmodified engines to use both petrol and the 24% blend. The modifications to run on E95 were a heated inlet manifold due to the cooling effect of ethanol, changes to the carburettor, the fuel tank and fuel line replaced by one in tin and cadmium brass. The fuel filter was changed to accommodate a higher fuel flow and the compression increased to 12.1 because of the higher octane rating. Changes were also needed to the valve housings and catalytic converter.

In the USA, the initial blend contained 10% ethanol (Gasohol) but more recently a blend containing 85% ethanol (E85) has been introduced and flexible fuel engines have been developed which can use either E85 or petrol. Ethanol is also used in the USA to increase the oxygen levels in petrol with an addition of 7.6% and as a replacement for MTBE in reformulated petrol. Ethanol contains 35% oxygen which increases combustion and therefore reduces particulate and NO_x emissions.

Large-scale Production of Ethanol from Biomass

The large-scale production of ethanol as a fuel started in Brazil in 1975, followed by the USA in 1978. The amount of ethanol produced in the world in 2001 was 22,540 billion l (Fig. 6.1); global production was dominated by Brazil and the USA. Brazilian production is based on the fermentation of sugar from sugarcane, whereas the USA used starch extracted from maize. By 2006, global production had increased to 51 billion l (13.5 billion gallons), which represents 4.6% of global petrol consumption. Bioethanol has the potential to replace 353 billion l of petrol, which is 32% of the global petrol consumption (1103 billion l) when used as 85% addition (E85) (Balat et al., 2008). In 2006, the USA produced 4.85 billion gallons (18.3×10^9 l) of bioethanol which has overtaken the production by Brazil of 4.49 billion gallons (16.97×10^9 l) (Fig. 6.2).

The plants installed in Brazil are small, in the region of 100,000 l in capacity, compared with the large units installed in the USA at around 0.8 billion l per year.

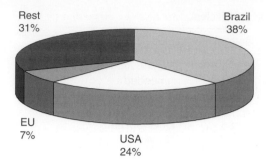

Rest
31%

Brazil
38%

EU
7%

USA
24%

Fig. 6.1. Distribution of production of ethanol worldwide in 2001, from a total of 22,540 billion l. (From Licht, 2006.)

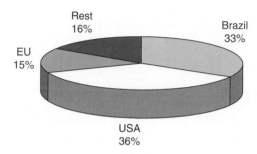

Rest
16%

Brazil
33%

EU
15%

USA
36%

Fig. 6.2. Distribution of ethanol production worldwide in 2006, from a total of 51,000 billion l. (From Balat *et al.*, 2007.)

Table 6.5. Leading ethanol producers in the USA 2006. (From Solomon *et al.*, 2007.)

Company	Capacity (10^9 l/year)
Archer Daniels Midland	4.1
VeraSun Energy	0.87
Hawkeye Renewables	0.84
Aventine Renewable Energy	0.57
Cargill Inc.	0.46
Abengoa Corp.	0.42
New Energy Corp.	0.38
Global Ethanol/Midwest	0.36
Total	19.0

Table 6.5 lists some of the largest ethanol producers in the USA, with the largest volume produced at 4.1 billion l (1.08 billion gallons). This difference is in part due to the USA plants being attached to very large maize-processing plants, whereas in Brazil locally grown sugarcane is processed in smaller units avoiding extensive transport.

Ethanol production process in the USA

In the case of sugarcane the sugar can be pressed from the cane and used directly in fermentation to produce ethanol. Starch on the other hand cannot be used by yeasts in fermentation and so has to be converted to glucose before it can be used. This is the main difference between the processes used in the USA and Brazil.

Figure 6.3 outlines the process that is used to obtain glucose from maize. The process was developed initially to produce starch from maize where the starch was used in food formulation, for the production of high fructose maize syrup, a low-calorie sweetener and a wide range of starch products. The harvested maize is soaked in water to loosen the kernels which are passed through a mill which removes the germ or embryo. The wash water is known as maize steep liquor and is one of the components of the medium used to grow penicillin. The separated germ is used for germ meal as a high protein supplement or pressed to extract maize oil (Mazola) used widely in cooking. The kernels, which now contain mainly starch, are ground, washed and centrifuged to remove fibre and gluten which are used in the food industry. What remains is starch which can be used in food and other products. Some of the starch is converted into glucose so that it can be processed into a mixture of glucose and fructose known as high fructose maize sugar (HFCS) for use as a low-calorie sweetener. The conversion of starch into glucose is carried out by starch-degrading enzymes, the amylases. Starch is synthesized in the chloroplast where glucose molecules are linked together with α-D-1,4 glycosidic linkages to form long chains (Fig. 6.4). At some stages, a branching by enzyme adds a side chain with a α-D-1, 6 linkage which gives starch a more rigid structure. The starch molecules can be broken down either by hydrolysis with acid or by enzymatic breakdown. The enzymatic

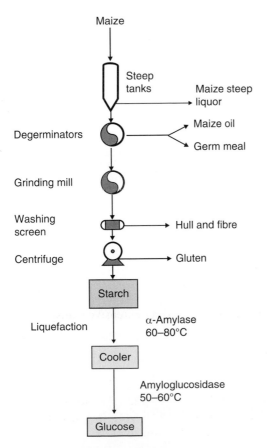

Fig. 6.3. The processing of maize in the USA for the production of glucose. (From Scragg, 2005.)

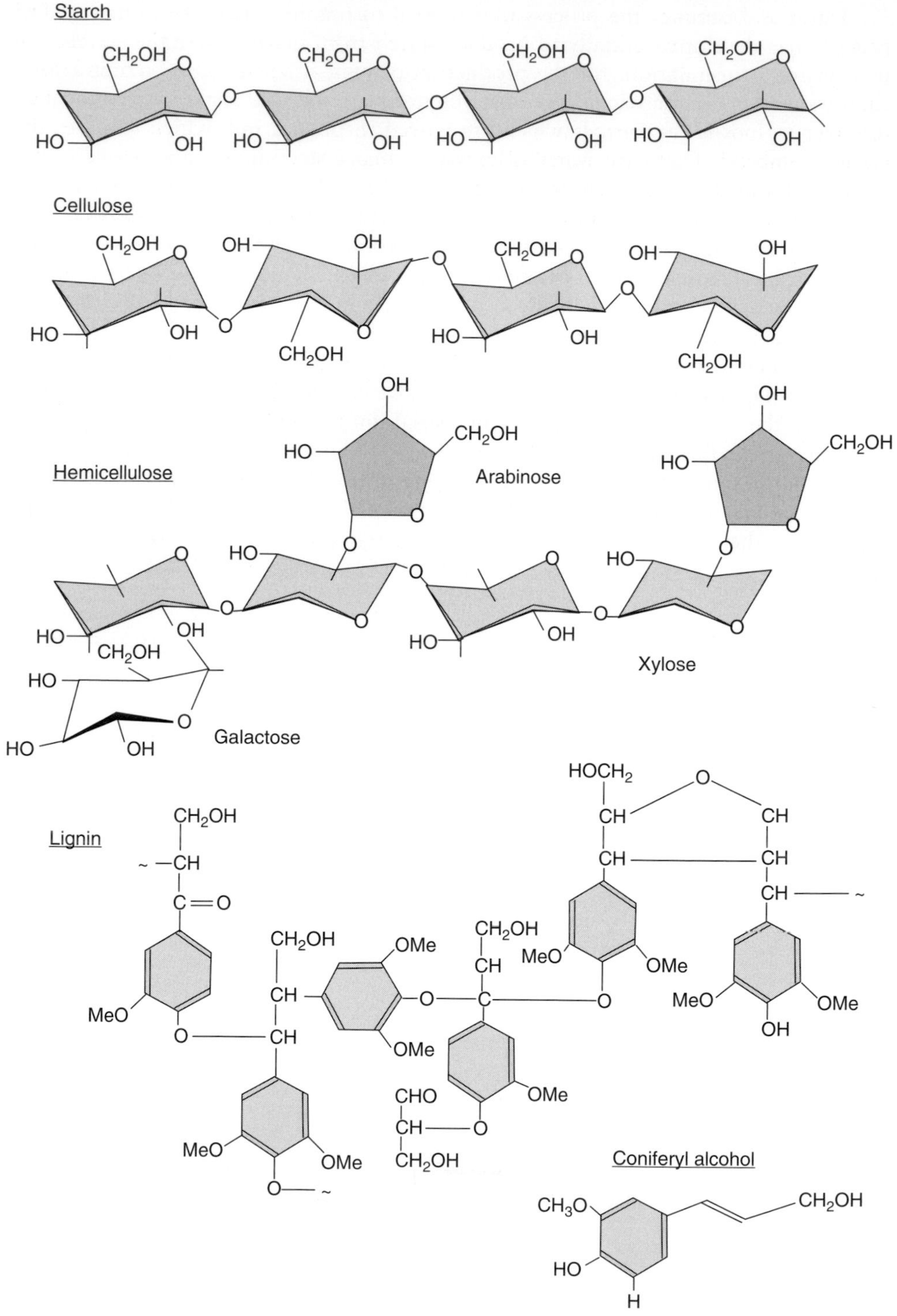

Fig. 6.4. The structure of starch, cellulose, hemicellulose and lignin.

process is normally used as it produces fewer by-products. The starch forms a stiff paste with water, which is first liquefied by heating to 60–80°C and the enzyme α-amylase added. The enzyme breaks the long glucose chains in the starch into shorter sections, known as long chain dextrins. In some cases, a high temperature α-amylase is used and the starch heated to 100°C. After a short period of time, the liquefied starch is cooled to 50–60°C and another enzyme, amyloglucosidase, added. This enzyme converts the dextrins into glucose.

Once glucose has been produced, it can be used as a substrate for ethanol fermentation by yeast. Figure 6.5 shows a typical process for the production of ethanol from glucose. The glucose from the starch is mixed with salts, nitrogen and phosphorus

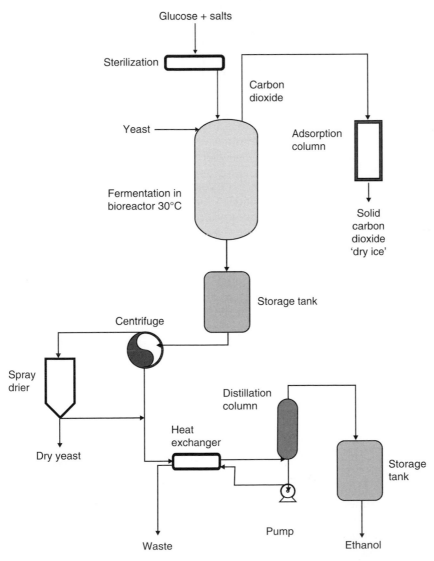

Fig. 6.5. A typical process for the production of ethanol from glucose produced from maize starch.

and the mixture sterilized by heating to 120°C for 2–5 min in a continuous sterilizer. Once sterilized, the medium is run into a large bioreactor (200,000 l and above), yeast added from a seed bioreactor (1–10% volume of the main reactor) and the culture incubated at 30°C for a few days. Carbon dioxide produced during the fermentation can be adsorbed and used to make solid carbon dioxide, 'dry ice'. Once fermentation has finished the yeast cells are removed by centrifugation and the medium, sometimes known as 'beer', is warmed by passing through a heat exchanger and then distilled. Distillation is needed to concentrate the ethanol and is the major energy-consuming stage. The fermentation yields about 10% ethanol and it needs to be more than 95% to be used as E95 or 100% if used as a blend. Heating a 10% ethanol solution will yield a vapour containing more ethanol than water and the remaining solution will contain more water so that a limited amount of enriched ethanol can be obtained. But with a series of distillations a concentration of 95.6% ethanol can be obtained. At 95.6% the liquid and vapour have the same concentration so no further concentration is possible. The mixture is known as an azeotrope. However, by using a distillation column separated by plates a series of separate distillations can be produced and this will give the azeotrope in one distillation. To produce anhydrous ethanol a second distillation is required where benzene is added and this on distillation gives pure ethanol and the benzene can be recovered and used again.

Ethanol production in Brazil

The production of ethanol is considerably simpler in Brazil as there is no starch to process. The sugarcane is harvested and milled to extract sugar (sucrose) and the rest of the plant, known as 'bagasse', is retained as it can be burnt in boilers. The sugar can be processed to produce sugar and the residue and molasses used for fermentation or the sugar juice used directly (Fig. 6.6). The sugar and salts are run into 100,000–400,000 l open bioreactors and inoculated with yeast. After fermentation has ceased, the yeast is removed by flocculation or centrifugation and the liquid distilled. If more than 95.6% ethanol is required a second distillation is carried out with the ethanol blended with fusel oil. The residue from the first distillation can be used as a fertilizer. The economy of the process is improved greatly as the residue from the sugarcane (bagasse) is used to fire boilers which supply steam for the distillation process.

Pathways Involved in Ethanol Metabolism

One of the reasons for substituting ethanol for petrol is that ethanol can be produced from biological material and hence is both renewable and sustainable, and reduces carbon emissions.

The ability of microorganisms to produce alcohol from sugars has been known since Egyptian times and could be regarded one of the first uses of biotechnology. The best-known and most widely used microorganism involved with the production of ethanol has been the yeast *Saccharomyces cerevisiae*, but it is not the only one. In the absence of oxygen, the yeast will switch its metabolism to fermentation producing ethanol and carbon dioxide. The lack of oxygen inhibits the citric acid cycle so pyruvate would be expected to accumulate (Fig. 6.7). However, under these conditions

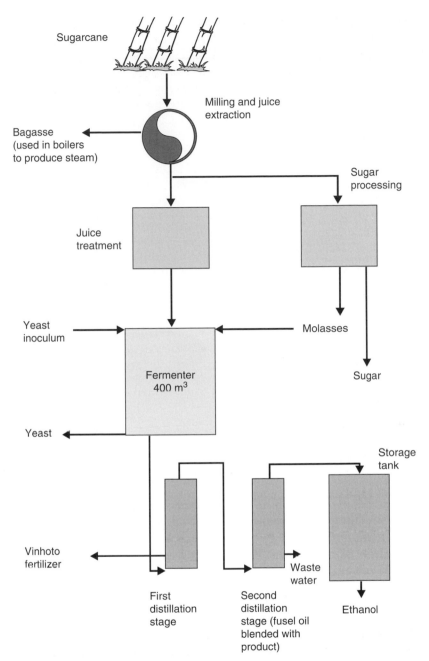

Fig. 6.6. Typical process for the production of ethanol from sugarcane in Brazil. (Modified from Poole and Towler, 1989.)

pyruvate is converted to acetaldehyde by the enzyme pyruvate decarboxylase with the release of carbon dioxide. Acetaldehyde is then converted to ethanol by the enzyme alcohol dehydrogenase. The overall equation is given below:

$$C_6H_{12}O_6 = 2C_2H_5OH + 2CO_2 \tag{6.1}$$

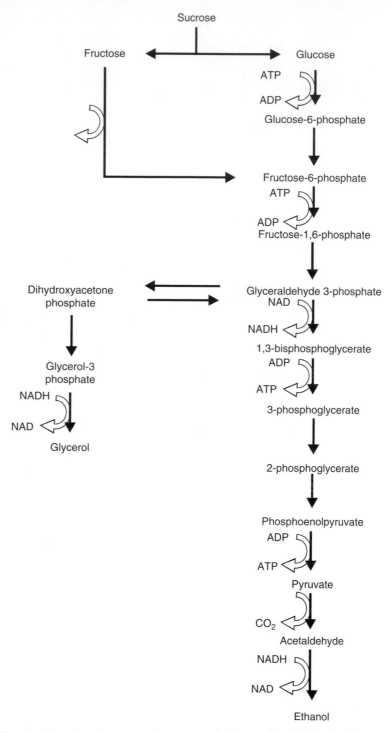

Fig. 6.7. The metabolism of sucrose, fructose and glucose by yeasts. Glycolysis yields pyruvate which is decarboxylated to acetaldehyde and acetaldehyde reduced to ethanol.

The theoretical yield of ethanol from this equation is 51% of the substrate added but some energy is required to maintain the cells so that the yield is about 95% of the theoretical yield with pure substrates. However, with industrial systems the best yields are around 91%. The concentration of ethanol obtained by fermentation is normally from 5 to 10% as ethanol begins to inhibit growth above 5%. Concentrations of 10% ethanol can be obtained with pure substrates. The reason for the loss of viability as the ethanol concentration increases is that ethanol is a solvent and disrupts the cells' lipid-protein membrane making it increasingly leaky. Yeast strains with a higher tolerance to ethanol have membranes containing a higher proportion of longer-chain unsaturated fatty acids. Higher concentrations of ethanol can be obtained using high concentrations of substrates, and ethanol tolerant (10–18%) strains but the process is much slower. The limitation of *S. cerevisiae* is that it cannot utilize pentoses such as xylose and arabinose and more complex carbohydrates like starch and cellulose.

Other ethanol-producing microorganisms

S. cerevisiae is not the only microorganism capable of producing ethanol (Table 6.6). Other yeasts, bacteria and fungi can also ferment sugars to produce ethanol. Ethanol-producing bacteria have been of commercial interest because they have a faster growth rate and can be easily genetically engineered. *Escherichia coli* uses a different pathway to produce ethanol from pyruvate (Fig. 6.8) with an enzyme pyruvate formate lyase forming acetyl-CoA, which is then reduced by two steps with alcohol dehydrogenase to ethanol.

Other ethanol-producing bacteria metabolize glucose via the Entner–Doudoroff pathway which consumes less ATP than glycolysis (Fig. 6.9). One of the best examples

Table 6.6. The substrates that can be utilized by ethanol-producing microorganisms.

Organism	Substrate utilized
Yeasts	
Saccharomyces cerevisiae	Glucose, fructose, galactose, maltose, maltotriose, xylulose
Saccharomyces carlsbergensis	Glucose, fructose, galactose, maltose, maltotriose, xylulose
Kluyveromyces fragilis	Glucose, galactose, lactose
Candida tropicalis	Glucose, xylose, xylulose
Bacteria	
Zymomonas mobilis	Glucose, fructose, sucrose, engineered to use xylose
Clostridium thermocellum	Glucose, cellobiose, cellulose
Escherichia coli	Xylose, no ethanol tolerance
Klebsiella oxytoca	Xylose, cellobiose, glucose
Clostridium acetobutylicum	Xylose to acetone and butanol, ethanol in small quantities
Lactobacillus plantarum	Uses cellobiose faster than glucose
Lactobacillus casei	Lactose, useful for whey utilization
Lactobacillus xylosus	Uses cellobiose if nutrients supplied, glucose, xylose and arabinose

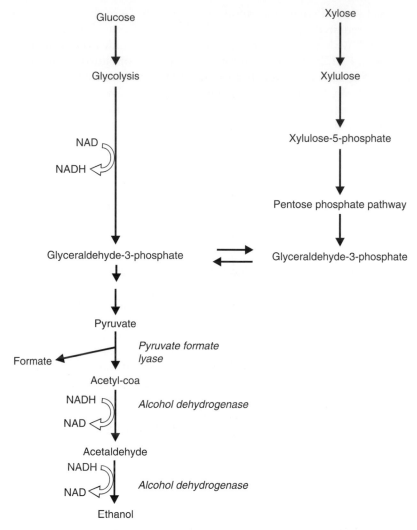

Fig. 6.8. The pathway that *Escherichia coli* uses to produce ethanol from glucose and xylose.

is the bacterium *Zymomonas mobilis* which has a higher growth rate, is more ethanol-tolerant than yeasts but still only metabolizes glucose, fructose and sucrose.

Substrates for Ethanol Production

At present the two principal substrates that have been used commercially to produce ethanol are sugar (sucrose) and starch. There are problems with both these substrates in terms of can these crops supply sufficient ethanol, and in doing so will they compromise the supply of food crops. There are other abundant, inexpensive, non-food substrates and the most obvious is lignocellulose.

Lignocellulose is the most abundant potential source for bioethanol production with a potential yield of 442 billion l (Balat *et al.*, 2007). For countries where biofuel

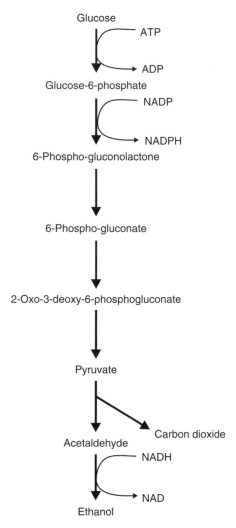

Fig. 6.9. The production of ethanol via the Entner–Doudoroff pathway.

crops are difficult to grow, lignocellulose is an attractive option. Lignocellulose can be obtained from trees species, wood residues, sawdust chips, construction residues, municipal wastes, paper, sewage sludge, maize stover, straws and grasses such as Miscanthus, switchgrass, sorghum, reed canary grass, bagasse, sugarbeet pulp, softwood, wheat straw, rice straw, pulp and paper mill residue, forest thinnings, municipal solid waste, winter cereals, and recycled paper.

Unfortunately, there are no ethanol-producing yeasts, other than genetically modified (GM) strains, that are capable of metabolizing starch and lignocellulose and only a few bacteria are capable of metabolizing lignocellulose. To make lignocellulose and starch suitable for fermentation both need to be converted into sugars in an inexpensive process.

Lignocellulose

Lignocellulose consists of three polymer types, cellulose, hemicellulose and lignin, which are the main constituents of plant cell walls (Fig. 6.4). The primary cell wall consists of cellulose fibres embedded in a polysaccharide matrix of hemicellulose and pectin. Cellulose is the most abundant plant compound and the second most abundant is lignin which provides mechanical support and protection in plants. The composition of some lignocellulose sources is given in Table 6.7 in terms of lignin, cellulose and hemicellulose. Cellulose is a polymer of glucose linked together by 1,4-glycosidic bonds. Hemicellulose is heterogeneous polymers with a backbone of 1,4-linked xylose residues but contains short side chains containing other sugars such as galactose, arabinose and mannose. Xylose is the predominant sugar in hardwoods and arabinose in agricultural residues.

Lignin is a highly branched polymer of phenyl-propanoid groups such as coniferyl alcohol (Fig. 6.4). Cellulose represents 40–50% of dry wood, hemicellulose 25–35% and lignin 20–40% depending on the plant type. Although lignocellulose is abundant, it cannot be metabolized by yeast and therefore needs to be broken down to its constituent sugars before it can be used. In addition not all yeast can metabolize the pentose sugars derived from the hemicellulose. The composition of agricultural lignocellulose sugars is shown in Table 6.8. *Saccharomyces cerevisiae* can ferment

Table 6.7. The composition of various biomass and waste materials. (From Hamelinck *et al.*, 2005; Ballesteros *et al.*, 2004; Champagne, 2007.)

Substrate	Cellulose	Hemicellulose	Lignin
Hardwood eucalyptus	49.5	13.1	27.7
Softwood pine	44.6	21.9	27.7
Switchgrass	32.0	25.2	18.1
Wheat straw	35.8	26.8	16.7
Cattle manure	27.4	12.2	13.0
Pig manure	13.2	21.9	4.1
Poultry manure	8.5	18.3	4.9

Table 6.8. Sugar composition as a percentage of some agricultural lignocellulose materials. (Adapted from van Maris *et al.*, 2006.)

Sugar	Maize stover	Wheat straw	Bagasse	Sugarbeet pulp	Switchgrass
Fermented by yeast					
Glucose	34.6	32.6	39.0	24.1	31.0
Mannose	0.4	0.3	0.4	4.6	0.2
Galactose	1.0	0.8	0.5	0.9	0.9
Not fermented					
Xylose	19.3	19.2	22.1	18.2	0.4
Arabinose	2.5	2.4	2.1	1.5	2.8
Uronic acids	3.2	2.2	2.2	20.7	1.2

glucose, mannose and galactose, but not the other sugars. Depending on the source of lignocellulose the sugar produced will change. Glucose and xylose are the main sugars in the agricultural lignocellulose except in switchgrass. Investigations are under way to find organisms which can ferment these other sugars and produce ethanol or to genetically manipulate yeasts to be able to metabolize these sugars.

Lignocellulose is difficult to break down into sugars, but a number of technologies are under investigation including enzymes. Because of the presence of hemicellulose and lignin and the crystalline nature of cellulose in lignocellulose some form of pretreatment is required before enzymatic or chemical hydrolysis. These pretreatments are shown in Fig. 6.10, and include carbon dioxide, steam and ammonia explosion, mechanical grinding, acid, white rot fungi treatment and ozonolysis.

Steam explosion

High-pressure and high-temperature steam can be used to treat the lignocellulose material where hemicellulose is hydrolysed by acids released during steam treatment. Acid addition increases the sugar yields but sulfuric acid can yield sulfur dioxide which can be inhibitory to further treatment. Steam treatment is less energy intensive than mechanical disruption.

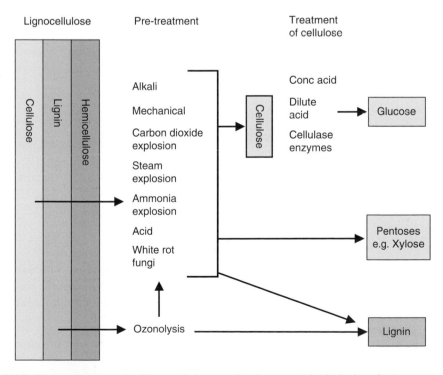

Fig. 6.10. The pretreatment of lignocellulose and subsequent hydrolysis prior to fermentation.

Mechanical disruption

The grinding of lignocellulose materials into small particles increases the surface area and allows subsequent enzyme or acid treatment to hydrolyse the cellulose. The process requires a considerable energy input and is not as effective as other treatments.

Ammonia explosion

The lignocellulose is milled and the ground lignocellulose, with a moisture content of 15–30%, is placed in a pressure vessel with ammonia (1–2 kg/kg biomass) at pressures of 12 atmospheres for 30 min. No sugars are released but the hemicellulose and cellulose are opened up to enzymatic digestion.

Acid treatment

Acid treatment can use sulfuric, hydrochloric, nitric and phosphoric acids although sulfuric is used most widely. Acid treatment converts the hemicellulose to sugar (xylose) (80–95%) and furfural, and increases cellulose digestibility. The treatment time is short (minutes), and depending on the substrate between 80 and 95% of the hemicellulose sugar can be recovered.

Alkaline treatment

Alkali treatment reduces the lignin and hemicellulose content and can be carried out at lower temperatures and pressures than acid treatment. The treatment also increases the surface area of the biomass.

Ozonolysis

Ozone can be used to degrade lignin and hemicellulose (Sun and Cheng, 2002). The advantages are it removes lignin, produces no toxic residues and it is carried out at room temperature and pressure.

Enzymes

Fungal enzymes from white and brown rot fungi, such as *Sporotrichum pulverulentum* and *Pleurotus osteatus*, can be used to pretreat lignocellulose. The brown rot fungal enzymes degrade cellulose whereas white rot fungal enzymes degrade lignocellulose.

Cellulose breakdown

After pretreatment the cellulose is suitable for hydrolysis to glucose and a number of methods can be used (Fig. 6.10).

Concentrated acid

Concentrated acid hydrolysis of cellulose gives a rapid and complete conversion to glucose using 70% sulfuric acid for 2–4 h. The problems are associated with the difficulties of handling concentrated sulfuric acid and the cost of the acid, which requires recovery and reuse to be economical.

Dilute acid hydrolysis

The cellulose is broken down by dilute acid in complex reaction at a higher temperature than the hemicellulose reaction and has a sugar recovery of around 50%.

Cellulase enzymes

Once the structure of the cellulose has been opened up by the pretreatment, enzymatic hydrolysis can proceed. The crude cellulase enzyme is a consortium of enzymes, which operate under mild conditions, pH 4.8 and 45–50°C. Although cellulase is commercially available it is usually obtained from fungi such as *Trichoderma reesei*, and the yields are better than acid hydrolysis. Cellulases can be produced by both fungi and bacteria which can be grown both aerobically, and anaerobically. The bacteria include *Clostridium*, *Cellulomonas*, *Bacillus*, *Thermomonospora*, *Bacteroides*, *Erwinia*, *Acetovibrio*, *Microbispora* and *Streptomyces*. Three enzymes are involved in the hydrolysis, endo-1,4-β-glucanases (endoglucanases), cellobiohydrolyases (exoglucanases), and β-glucosidases. The endoglucanases cleave the cellulose chain randomly and the exoglucanases hydrolyse the cellulose chain, releasing glucose and cellobiose. The β-glucosidases catalyse the conversion of cellobiose to glucose.

The fungi produce all three types of cellulase but the exoglucanases are the major enzymes with *T. reesei*. The *Trichoderma* sp. are considered the best of cellulase enzyme producers. Cellulose on hydrolysis liberates cellobiose which is cleaved into two molecules of glucose by the enzyme β-glucosidase. The disadvantage of the enzyme process is that both products glucose and cellobiose act as inhibitors of cellulase and β-glucosidase enzymes.

Types of Processes Used for Lignocellulose Breakdown

Once the hemicellulose and lignin have been removed ethanol production from cellulose and is often carried out in two stages, first the hydrolysis of cellulosic material to sugars by cellulase enzymes, and second the fermentation of the sugars. The rate of cellulose hydrolysis is about half the speed of the production of ethanol by yeast. A number of process formats have been developed.

Separate hydrolysis and fermentation (SHF)

In this case the hydrolysis of cellulose is carried out separately from the fermentation but if the products of hemicellulose are to be included a second bioreactor is used to ferment pentose sugars such as xylose. The use of a separate bioreactor gives a higher yield of ethanol and less energy is required. Considerable efforts have been made to improve the yield of ethanol using the normal yeast fermentation. A number of process changes have been investigated in order to improve the economics of ethanol production. The traditional method of fermentation has been batch culture in a non-stirred cylindro-conical vessel where the sugar and salts are inoculated with yeast and the fermentation allowed to proceed until the sugar is exhausted. Other forms of bioreactor operations and designs have been investigated in order to improve ethanol productivity, as this affects the cost of the final product. The fermenter can be operated in a batch-fed mode where batches of fresh medium are added at times during the fermentation. This avoids substrate inhibition where a high substrate concentration at the beginning of the fermentation may inhibit growth. In the continuous mode medium is added continuously throughout the fermentation and cells and medium removed at the same rate to keep the volume in the fermenter the same. This allows the operator to run the fermenter for a long period without having to waste time cleaning, refilling and sterilizing the fermenter. Fermenters of various designs where cell recycling has been used have considerably higher productivity which is due to maintaining a high cell density throughout the process.

The traditional fermentation vessel is not stirred but stirred tanks can be used to give good mixing and a more rapid growth rate. Alternative fermenter designs have been tested in order to improve the rate of growth and ethanol production. The tower fermenter is just an elongated tank with a high aspect ratio. The fluidized bed fermenter operates by mixing the cells by pumping the medium up through the base of the fermenter, thus fluidizing the cell mass at the bottom of the tank. A membrane fermenter keeps the cells separate from the medium with a semi-permeable membrane. This allows the fermenter to retain a high cell density and thus a higher rate of ethanol production. Examples of the productivity of these various systems are given in Table 6.9.

Another method for maintaining a high cell density is to immobilize the cells on or in some form of support. This retains the cells within the bioreactor at a high density and allows for a continuous process. Examples of immobilized cells are given in Table 6.9. It is difficult to compare results as the glucose used differs in concentration but it is clear that cell recycling results in increased productivity.

Table 6.9. Ethanol production using fermenters of different designs and operation.

Bioreactor	Substrate glucose (g/l)	Ethanol formed (g/l)	Ethanol productivity (g/l/h)
Stirred tank	100	–	7.0
Stirred tank with cell recycling	100	–	29.0
Tower	112	51.4	26.5
Fluidized bed	150	–	40.0
Membrane bioreactor with cell recycling	100	50	100.0
Immobilized cells in packed bed	196	93.5	36.5
Immobilized cells in cross-flow bed	103	48	37.1

Simultaneous saccharification and fermentation (SSF)

One of the most important advances in ethanol production was the development of simultaneous saccharification and fermentation (SSF). In this system, yeast ferments the glucose produced by the cellulase enzymes in the same vessel and at the same time. The cellulase enzymes therefore do not suffer from feedback inhibition from their products glucose and cellobiose as the fermentation removes these inhibitors. This increases hydrolysis rates, reduces enzyme levels, shortens process time and requires smaller bioreactor volumes. The drawbacks to the system are the differences in the optimal conditions for the enzymes and yeasts which reduces the process rate. The cellulase normally operates at 40–50°C, whereas yeast fermentation is carried out at 30°C. One way of avoiding this is to use thermotolerant yeasts like *Kluyveromyces marxianus*. The use of yeasts that can assimilate pentoses, for example *Candida acidothermophilum*, *C. brassicae* and *Hansenula polymorpha*, would also improve the process.

The hydrolysis can be carried out by an enzyme mixture or enzyme-producing microorganisms. The organisms used in SSF are often *T. reesei* which provides the enzymes and *S. cerevisiae* for the fermentation, run at a temperature of 38°C. The temperature is a compromise between yeast optimum of 30°C and the hydrolysis optimum of 45–50°C. The major advantages of SSF have been found to be increase in hydrolysis due to the reduction in feedback inhibition, lower enzyme or organism requirement, higher product yield, less contamination as sugar levels are kept low, shorter process time and smaller bioreactor volumes.

Simultaneous saccharification and co-fermentation (SSCF)

In simultaneous saccharification and co-fermentation (SSCF) technology enzymatic hydrolysis of lignocellulose continuously releases hexose (glucose) and pentose (xylose) sugars and two organisms jointly assimilate the pentoses and hexoses. Examples of the organisms are *Pichia stipitis* and *Brettanomyces clausennii* and in other cases recombinant microorganisms have been used.

Consolidated bioprocessing

In this process, fermentation and enzyme production are carried out by a single microorganism growing in a single bioreactor. There are no microorganisms which can carry out this process at present but some are under development.

Extraction of Ethanol

As ethanol builds up in the fermenter it will begin to inhibit the growth of many microorganisms, and some microorganisms are very sensitive to ethanol (Cardona and Sanchez, 2007). To ameliorate this inhibition four methods for the removal of ethanol during fermentation have been used: vacuum, gas stripping, membranes and liquid extraction (Figs 6.11 and 6.12).

Ethanol has a boiling point of 78°C so that applying a vacuum to the fermenter will remove ethanol from the medium. This technique has not been widely used but in

(a) Vacuum fermentation

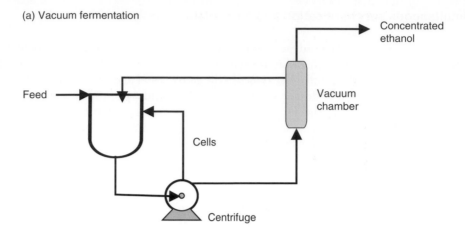

(b) Fermentation with gas stripping

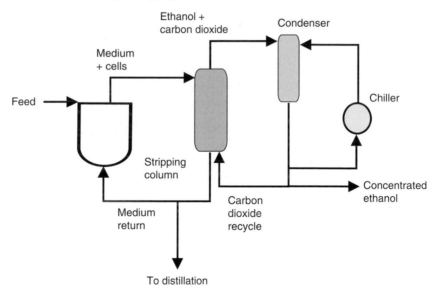

Fig. 6.11. Ethanol removal from fermentation by (a) vacuum and (b) gas stripping. (Redrawn from Cardona and Sanchez, 2007.)

one case a 12-fold increase in ethanol production was obtained. Gas stripping involves passing a gas, in many cases carbon dioxide, through the culture in a separate column to sweep out the ethanol (Fig. 6.11b). More complex arrangements have been designed and pilot plant units have been operated for some time. The use of ceramic membranes for the separation of ethanol from the growth medium has been investigated using a separate unit from the fermenter known as a pervaporation unit. The use of membranes to immobilize cells can also be combined with ethanol removal. It has been shown that pervaporation can reduce the cost of ethanol by 75% as much of the cost is associated with distillation.

Ethanol can be extracted from the fermentation medium by using a bio-compatable solvent such as aliphatic alcohols *n*-dodecanol, oleyl alcohol, and dibutyl

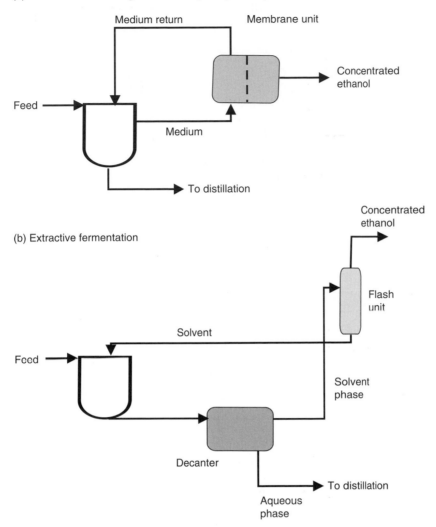

(a) Ethanol removal using membranes (pervaporation)

Medium return

Membrane unit

Concentrated ethanol

Feed

Medium

To distillation

(b) Extractive fermentation

Concentrated ethanol

Flash unit

Solvent

Feed

Solvent phase

Decanter

To distillation

Aqueous phase

Fig. 6.12. The removal of ethanol by (a) membrane (pervaporation) and (b) liquid extraction. (From Cardona and Sanchez, 2007.)

phthalate. The solvent can be added to the fermentation and the aqueous and water phases can be separated in a decanter. In a version of this technique a two-phase fermentation can be run using two incompatible polymers and the ethanol can be partitioned into one phase and the biomass in the other.

Commercial Lignocellulose Processes

One of the problems of using enzymes to degrade lignocellulose is the cost of the enzymes. Recent reduction in the cost of cellulose enzymes (Greer, 2005) has allowed lignocellulose to be considered as a viable alternative to sugar and starch as a substrate

Table 6.10. Some of the pilot plants constructed for the production of ethanol from lignocellulose. (From Solomon *et al.*, 2007.)

Company	Location	Feedstock	Capacity 10^2 kg/day	Start date
Iogen	Ottawa, Canada	Wood chips	9.0	1985
Iogen	Ottawa, Canada	Wheat straw	9.0	1993
Masada/TVA	Muscle Shoals, USA	Wood	Na	1993
SunOpta	Norval, Canada	Various	4.5	1995
Arkenol	Orange, USA	Various	9.0	1995
NREL/DOE	Golden, USA	Maize stover	9.0	2001
Pearson Technologies	Aberdeen, USA	Wood, rice straw	0.27	2001
NEDO	Izumi, Japan	Wood chips	3.0	2002
Dedini	Pirassununga, Brazil	Bagasse	42.0	2002
Tsukishima Kikai Co.	Ichikawa, Japan	Wood residues	9.0	2003
Etek EtanolTeknik	Ornskoldvik, Sweden	Spruce sawdust	5.0	2004
PureVision	Ft Lupton, USA	Maize stover	9.0	2004
Universal Entech	Phoenix, USA	MSW	1.0	2004
Sicco A/S	Odense, Denmark	Wheat straw	1.0	2005
Abengoa Bioenergy	York, USA	Maize stover	52.0	2006

MSW, municipal solid waste.

for ethanol production. A number of countries have pilot plants in operation processing lignocellulose, although most are in the USA (Table 6.10) and larger commercial plants are being developed. Figure 6.13 shows the process developed by Iogen, a Canadian enzyme manufacturer, for the use of lignocellulose to produce ethanol. Wood chips or straw are treated with dilute acid steam explosion and the cellulose and hemicellulose hydrolysed with enzymes produced by *Trichoderma* sp. Initially the fermentation used *S. cerevisiae* but later a recombinant *Z. mobilis* has been introduced which could ferment the pentoses produced from the hemicellulose. The lignin is separated and burnt in a combined heat and power unit to provide energy for the process.

Another process is that designed by the National Renewable Energy Laboratory (NREL) (Fig. 6.14). The biomass, maize stover, is pretreated with dilute acid followed by simultaneous saccharification and co-fermentation (SSCF) with a recombinant *Z. mobilis* capable of fermenting glucose, xylose and *T. reesei*. The cellulose hydrolysis is carried out with *T. reesei* cellulases and the hemicellulase hydrolysate is detoxified before adding to the fermenter.

Biological Production of Ethanol from Synthesis Gas

It has been known for some time that some anaerobic bacteria can convert carbon monoxide, carbon dioxide and hydrogen into a mixture of ethanol, butanol and acetic acid. The pathway is shown in Fig. 6.15 where acetyl-CoA is produced via the acetogenic pathway (Henstra *et al.*, 2007). Synthesis gas is a mixture of carbon monoxide and hydrogen produced by the gasification of coal, biomass and wastes. In the first step the oxidation of carbon monoxide and hydrogen yields protons which are used to reduce carbon dioxide to formate. The formate bonds to tetrahydrofolate (THF), and this complex adds more protons finishing with a methyl-THF complex.

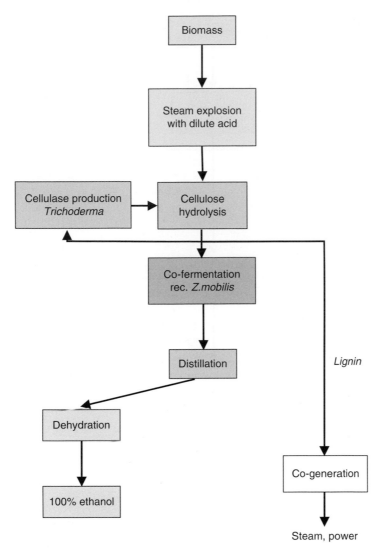

Fig. 6.13. The Iogen process for producing ethanol from lignocellulose. (Redrawn from Cardona and Sanchez, 2007.)

This complex reacts with carbon monoxide and CoA to produce acetyl-CoA cata-lysed by acetyl-CoA/carbon monoxide dehydrogenase. If insufficient carbon mon-oxide is available it can be produced from carbon dioxide. The production of acetyl-CoA has a negative energy balance which is recovered by the reduction of acetate formed from acetyl-CoA to ethanol. Butanol is formed from acetoacetyl-CoA. The overall balance in the formation of ethanol from syngas with an equimolar mixture of carbon monoxide and hydrogen is two-thirds of the carbon monoxide can be con-verted into ethanol (Rajagopalan *et al.*, 2002).

The bacteria that have been demonstrated to be capable of growth on hydrogen and carbon monoxide include both mesophilic and thermophilic bacteria and Archaea. Examples of the mesophiles are *Clostridium autoethanogenum* and *Eubacterium*

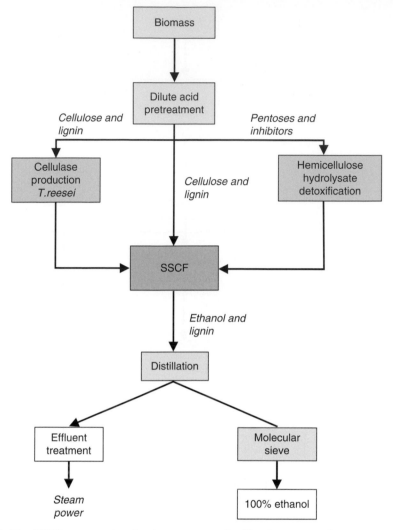

Fig. 6.14. The NREL system for simultaneous saccharification and co-fermentation (SSCF). (Redrawn from Cardona and Sanchez, 2007.)

methylotrophicum; thermophiles are *Moorella thermoacetica*, Archaea, and *Methanosarcina acetivorans*.

Recently the anaerobic fermentation of syngas has been developed from a laboratory process to a commercial process by Bioengineering Resources Inc. (BRI, 2007). An outline of the process is given in Fig. 6.16. A single module will process 100,000 t of biomass producing 6–8 million gallons of ethanol (US) and 5–6 MW of energy per year. The energy is derived from heat recovered from the gasifier as the gas has to be cooled to around 37°C before it is introduced into the fermenter and the exhaust gases from the fermenter. Distillation gives 95% ethanol and a molecular sieve is used to separate the remaining 5% water. There are no details of whether this process is more efficient and economical than the conversion of lignocellulose biomass into sugar or the Fischer-Tropsch process.

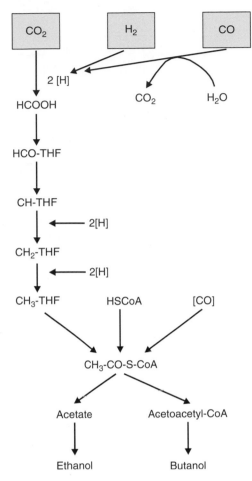

Fig. 6.15. Reductive acetyl-CoA pathway. THF, tetrahydrofolate. (From Henstra *et al.*, 2007.)

Butanol

Butanol is another alcohol which has been considered as liquid fuel as it has similar properties to ethanol (Table 6.2) but has a higher energy content. Butanol will give a higher mileage and can be mixed at any proportion with petrol. Butanol has been used as an industrial solvent, paint thinner and a component of brake fluids. It is less corrosive than bioethanol and can be transported through existing pipeline whereas ethanol has to be carried in tankers, by rail or on barges. It is also safer as it has a higher flash and boiling point. With all these advantages over ethanol it is perhaps not surprising that a number of ethanol plants have switched to butanol. BP, DuPont and British Sugar's plant in Wissington is being converted from ethanol to butanol and Virgin Fuels are interested in butanol produced from cellulose.

At present much of the butanol used is produced from petrochemicals but there is a renewable method of producing butanol using microorganisms. The biological production of butanol was first observed in 1861 when Pasteur isolated a butyric acid producing bacterium. Studies also showed that acetone was also formed and the organism was

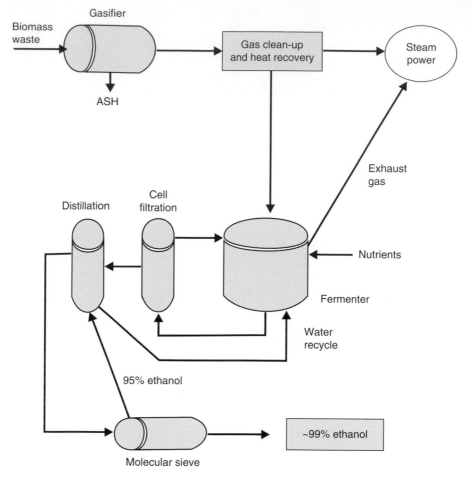

Fig. 6.16. An outline of the production of ethanol from biomass via gasification. (Redrawn from BRI, 2007.)

not able to grow in the presence of air. The industrial production of acetone and butanol by fermentation has a long history that started in 1914. Acetone and butanol were some of the first biotechnological products and the process that developed was one of the largest. Before 1914 acetone was produced by heating (dry distillation) calcium acetate. Calcium acetate was produced by the dry distillation or pyrolysis of wood. The wood distillate contained about 10% acetic acid that was either distilled off into calcium hydroxide to form calcium acetate or directly neutralized with lime. Between 80 and 100 t of wood was required to produce 1 t of acetone. In 1910 Chaim Weizmann had been working in Manchester as part of a group working for Strange and Graham Ltd trying to produce butanol by fermentation. Butanol was needed as it could be used to form butadiene, a precursor of synthetic rubber. At the time natural rubber was in short supply, as Brazil was the only source and they did not allow the export of rubber trees from their country. By 1914 Weizmann and co-workers had isolated an anaerobic organism which was later named as *Clostridium acetobutylicum* that produced both acetone and butanol when grown on starch. In 1914, at the start of the First

World War, the demand for acetone increased rapidly as acetone was used as a solvent for nitrocellulose in the manufacture of cordite, a smokeless explosive. By 1915 the demand had exceeded supply and the Nobel Company approached Weizmann and the process of biological production of acetone was adopted rapidly. Brewing capacity was commandeered and by 1916 the bioreactor capacity had reached $700\,m^3$.

At the end of the First World War the demand for acetone reduced but butanol was still in demand as a solvent for the nitrocellulose paints used in the rapidly developing motor industry. Acetone was also being used as a solvent in the production of aircraft dopes and for the production of textiles and isoprene. Only certain *Clostridia* are capable of producing reasonable levels of acetone and butanol and *C. acetobutylicum* has been the one most studied and used in industrial processes. *C. acetobutylicum* is a gram-positive anaerobic spore-forming rod 0.6–0.9 μm wide and 2.4–4.7 μm long. It is motile and will ferment arabinose, galactinol, fructose, galactose, glucose, glycogen, lactose, maltose, mannose, salicin, starch, sucrose, trehalose and xylose. The optimum growth temperature is 37°C. As the bacterium will form spores readily when the nutrients are exhausted it can be easily maintained as spores mixed with sterile soil. Loss of solvent-forming potential is a common problem with *C. acetobutylicum* cultures but heat treatment restores solvent-forming ability. The concentration of substrate normally used was 6.0–6.5% and the maximum yield of solvent formed was 37% of the substrate used. However, in practice the yields are around 30% with a ratio of butanol/acetone/ethanol of 6:3:1 with small amounts of hydrogen and carbon dioxide being formed as well. Thus 100 t of substrate will yield about 22 t of butanol. The yields depend upon a number of factors including the strain of microorganism, temperature, pH and substrate. In the 1930s, a bacterium *C. saccharobutylicum* was isolated which when grown on sucrose formed acetone and butanol only.

During the exponential phase little solvent is produced but butyric and acetic acids were formed causing the pH of the medium to drop from 6.0 to below 5.5 (Fig. 6.17). In the stationary phase the accumulation of acetone, butanol and ethanol

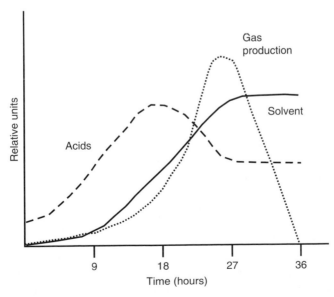

Fig. 6.17. The anaerobic production of acetone and butanol.

proceeds rapidly at the expense of the acids and therefore the pH rises. In culture *C. acetobutylicum* can be in three states: acidogenic where acetic and butyric acids are formed at neutral pH; solventogenic where acetone, butanol and ethanol are formed at low pH; and alcohologenic where butanol and ethanol are formed but no acetone at neutral pH, so that it is important to monitor or maintain pH.

Although molasses-based fermentations were more economical than the original starch substrate the expansion of the petrochemical industry from 1945 onwards meant that by the 1960s the process had ceased to be used. The reasons for the decline of the acetone/butanol process were:

- Low yield of solvents (30–35% of substrate).
- Low solvent concentration in medium due to the toxicity of butanol and ethanol at 20–25 g/l.
- Phage sensitivity.
- Autolysin-induced autolysis in stationary phase.
- Cost of distillation.
- Production of considerable amounts of waste.
- High cost of molasses.
- Petrochemical production was cheaper.

However, since the late 1990s the process has been reevaluated in the light of modern developments in genetic manipulation and waste treatment and the sudden increase in oil prices in 1973 (Durre, 1998). The reasons for the possible re-introduction are:

- The process uses renewable substrates.
- Butanol can replace ethanol as a liquid fuel.
- The newer strains can grow on waste starch and whey and metabolic engineering is being attempted so that it can be grown on cellulose.
- The waste can now be treated anaerobically forming biogas.
- The process may be able to operate at 60°C so that the solvents can be removed as they are formed.
- Solvent may be recovered during fermentation using reverse osmosis, perstraction, pervaporation, membrane evaporation, liquid–liquid extraction, adsorption and gas stripping (Durre, 1998). Any process that avoids distillation will be considerably cheaper and able to compete with fossil fuels.

It will be interesting to see how ethanol and butanol develop as liquid fuels in the EU and UK as in the short term much of the ethanol will have to be imported from Brazil.

Conclusions

At present the infrastructure is in place to use liquid fuels and therefore the replacement or addition to petrol will be ethanol at least in the short term. The main problem with ethanol is that when it is produced from starch considerable processing is required, which means a substantial input of energy. The most economical method is to make ethanol using sugar from sugarcane as in Brazil. However, the quantity of sugar needed to supply the volumes needed to replace ethanol may start to compromise the rainforest in Brazil as more and more land is use to grow sugarcane. The drivers for ethanol use are the directives in the EU and the USA to include ethanol in all petrol sold. If the amount of ethanol added to petrol is increased the increase in sugar and starch crops used for ethanol production may compromise food crops.

It is therefore important that ethanol production from the more abundant lignocellulose becomes industrial. Lignocellulose requires treatment and enzymatic degradation before it can be converted into ethanol. This processing requires energy and increases costs and these need to be reduced before lignocellulose ethanol can compete. It may be that lignocellulose will not be able to compete with Fischer-Tropsch fuels from biomass and wastes. Butanol may also supersede ethanol as the liquid fuel of choice.

7   Liquid Biofuels to Replace Diesel

Introduction

Both transport and industry rely heavily on the diesel engine that is widely used to power lorries, trains, tractors, ships, pumps and generators. The USA uses 50 billion gallons (1 gallon = 3.8 l) annually (Louwrier, 1998) and the consumption in the UK was 23.9 million t (10^6) in 2006 (IEA, 2008). The engine designed by Diesel ran for the first time on 10 August 1893, and the patent when filed proposed that the fuel could be powdered coal, groundnut oil, castor oil or a petroleum-based fuel (Shay, 1993; Machacon *et al.*, 2001). At this time, the growing petrochemical industry provided the best fuel, a crude oil fraction, now called diesel, which has been the fuel of choice for diesel engines ever since this time. Conventional diesel is produced by the distillation of crude oil and collecting middle distillate fractions in the range of 175–370°C. The fuel contains hydrocarbons such as paraffins, naphthenes, olefins and aromatics containing from 15 to 20 carbon molecules. To replace diesel without modifying the engine, any substitute will have to be similar to diesel in the following properties:

- A calorific value of 38–40 MJ/kg is a measure of the energy available in the fuel.
- A cetane number of around 50 is a measure of the ignition quality of the fuel.
- The viscosity of the fuel is important as it affects the flow of the fuel through pipelines and injector nozzles where a high viscosity can cause poor atomization in the engine cylinder.
- The flash point is a measure of the volatile content of the fuel and gives a measure of the safety of the fuel. The flash point for diesel is 64–80°C.
- It must be obtained from renewable resources such as biomass, oil crops and waste.
- It must be available in large quantities. For example the current use of diesel in the UK is 23,989,000 t where a 5% addition (on an energy basis) requires an addition of 5.75% by volume, which is equal to 1,199,450 t (1499 million l).

There are a number of possible sources of diesel replacements produced from agricultural products or microbial cultures, which are first-, second- and third-generation biofuels (Fig. 7.1). Some of the sources are as follows:

- Long-chain hydrocarbons (C 30) extracted from herbaceous plants, which can be cracked to form diesel, is a first-generation biofuel.
- Long-chain hydrocarbons (C 30) accumulated by some microalgae, which can also be cracked to form diesel, is a first-generation biofuel.
- Pyrolysis of biomass or waste to form bio-oil, which can be converted to diesel, is a second-generation biofuel.
- Gasification of biomass followed by Fischer-Tropsch synthesis of diesel (FT diesel) is a second-generation biofuel.

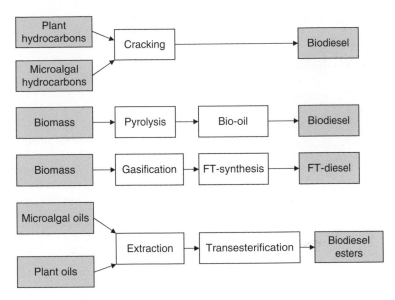

Fig. 7.1. The routes to the production of alternative diesels capable of replacing fossil fuel diesel.

- Transesterification of plant, animal and waste oils and fats to methyl esters (biodiesel) is a first-generation biofuel.
- Oil accumulated by some microalgae, extracted and transesterified into biodiesel, is a third-generation biofuel.

Synthetic Diesel, FT Synthesis

The Fischer-Tropsch (FT) synthesis was developed in the 1930s, by which a gas containing carbon monoxide (CO) and hydrogen (H_2) can be converted into long-chain hydrocarbons which have properties similar to crude oil products. A gas containing as its main components H_2 and CO can be produced by the high-temperature gasification of coal, biomass and waste, and is known as syngas. The gasification process produces a mixture of CO, H_2, methane (CH_4), carbon dioxide (CO_2), nitrogen (N_2) and water (H_2O). Natural gas can also be used in the FT process. The FT synthesis was used to produce diesel and petrol in World War II using coal as the starting material (Prins *et al.*, 2004). At present, syngas is mainly used by the chemical industry (Fig. 7.2) for ammonia production and only 8% is being used to produce hydrocarbon-based fuels called 'gas to liquid' (GTL) fuels where natural gas is used. In order to make the process sustainable, coal and gas should be replaced with biomass and waste materials. However, the process is costly and so this fuel is still under development.

Syngas production

At present, there are two industrial methods of producing syngas from biomass: a fluidized bed gasifier and entrained flow gasifier. The fluidized bed gasifier converts

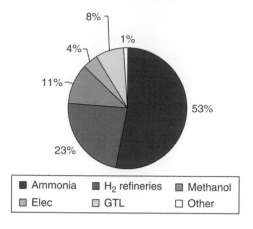

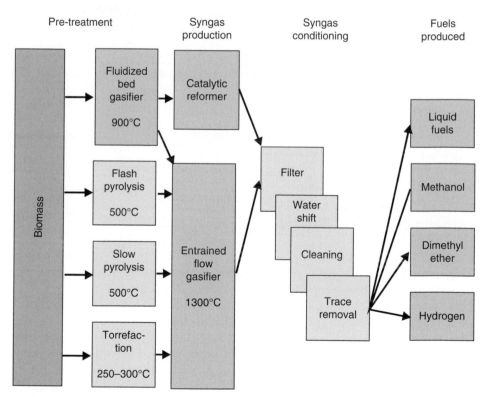

Fig. 7.2. Present industrial uses of syngas: ammonia production, hydrogen for refineries, methanol production, electricity generation and GTL which is the conversion of gas to a liquid fuel. (From van der Drift and Boerringter, 2006.)

Fig. 7.3. The processes that can be used to prepare syngas made from biomass for the Fischer-Tropsch synthesis of fuels. (From van der Drift and Boerrigter, 2006.)

biomass using an air-blown circulating fluidized bed operating at 900°C, but as the gas formed is not clean the system requires a catalytic reformer to remove many of the contaminants (Fig. 7.3). The gas from the fluidized bed gasifier contains H_2, CO, CO_2, H_2O and considerable amounts of hydrocarbons such as CH_4, benzene and tars (Table 7.1). The second option is entrained flow gasification where higher temperatures (1300°C) are used. This system requires a supply of very small particles to burn

Table 7.1. Typical gas composition produced by a fluidized bed gasifier using biomass. (Adapted from van der Drift and Boerrigter, 2006.)

Main constituents	Vol % dry wt	Lower heating value (LHV%)
Carbon monoxide (CO)	18	27.8
Hydrogen (H_2)	16	21.1
Carbon dioxide (CO_2)	16	–
Water (H_2O)	13	–
Nitrogen (N_2)	42	–
Methane (CH_4)	5.5	24.1
Acetylene (C_2H_2)	0.05	0.4
Ethylene (C_2H_4)	1.7	12.4
Ethane (C_2H_6)	0.1	0.8
BTX	0.53	10.5
Tars (total)	0.12	2.8

BTX, benzene, toluene, xylenes.

correctly so that any material used has to be milled, which is energy-intensive and makes handling difficult.

No matter which method is used to produce the gas, extensive syngas cleaning and conditioning are required before the FT process can be used to produce liquid fuels as the contaminates inhibit the catalyst. The syngas also needs to have a H_2/CO ratio of 2:1. The concentration of CO and H_2 can be adjusted in the water shift reactor which converts CO to H_2 and CO_2. The reverse can also be carried out as the syngas composition varies depending on the feedstock.

Forward (<250°C)

$$CO + H_2O = CO_2 + H_2 \qquad (7.1)$$

Backward (>500°C)

$$H_2 + CO_2 = CO + H_2O \qquad (7.2)$$

Fischer-Tropsch process

Figure 7.3 gives an overall view of the methods that can be used to produce biofuels from coal and biomass, and Table 7.2 gives the maximum concentration of impurities that syngas should have in order to be suitable for FT synthesis. Too high a concentration of impurities will poison the cobalt catalyst in the process.

The exothermic FT synthesis combines H_2 and CO when passed over a cobalt catalyst at a temperature of around 260°C producing a mixture of hydrocarbons including petrol (C_8—C_{11}) with an average of C_8H_{18} and diesel (C_{11}—C_{21}) with an overall hydrocarbon average of $C_{16}H_{34}$.

$$2H_2 + CO = CH_2 + H_2O \qquad (7.3)$$

The FT synthesis unit operations are given in Fig. 7.4 when using dried biomass. The dried biomass is gasified in an entrained flow gasifier at 900–1300°C in the presence

Table 7.2. Maximum concentrations for impurities allowed in syngas. (Adapted from van der Drift and Boerrigter, 2006.)

Impurity	Specification
H_2S, COS, CS_2	<1 ppmv
NH_3, HCN	<1 ppmv
HCL, HBr, HF	<10 ppbv
Alkali metals (Na, K)	<10 ppbv
Soot, ash	Complete removal
Organic (tar)	Not condensing
Hetero-organic components (S,N,O)	<1 ppmv

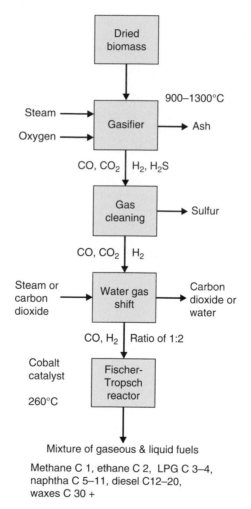

Fig. 7.4. Outline of the Fischer-Tropsch process using biomass to produce a mixture of hydrocarbons including petrol and diesel.

of steam and oxygen. In some cases, the biomass may be pretreated by pyrolysis or torrefaction (Fig. 7.3) or even taken from a fluidized bed gasifier. The ash is removed and the gas is cleaned of sulfur-containing compounds, and then the CO/H$_2$ ratio is adjusted by the water shift reaction. The cleaned gas is then passed over a cobalt catalyst in the FT reactor producing a range of hydrocarbons from CH$_4$ to waxes. The alpha factor shown in Fig. 7.5 describes the proportion of the various products formed, and this is affected by the catalyst used and process conditions. Maximum diesel production is around 30% of the total products at an alpha value of 0.85–0.9. The lower-temperature conditions which favour diesel production are 260°C, with cobalt-based catalyst at a pressure of 15–40 bar.

The process of gasification, gas cleaning and FT synthesis is a complex chemical process where the larger the scale, the more economic the process (Fig. 7.6). As the size increases, the conversion costs reduce, levelling out at around 1800 MWth (megawatts thermal) while the other costs remain static.

Thus, the production plant, using biomass to produce syngas and FT products, will be much larger compared to other biomass processes because of the increased efficiency

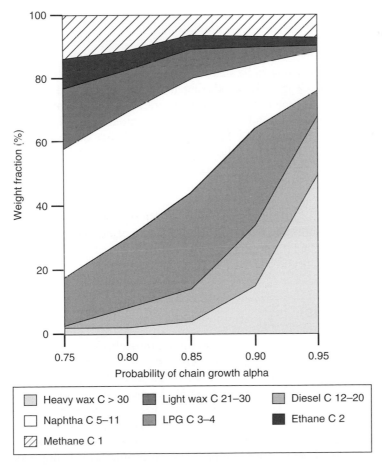

Fig. 7.5. The effect on the products formed in the Fischer-Tropsch process of the alpha factor, the probability of chain growth.

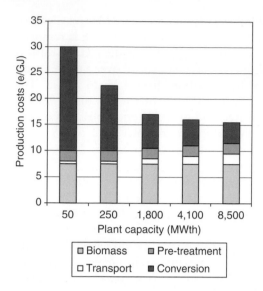

Fig. 7.6. The effect of scale on the economics of the Fischer-Tropsch process. (Redrawn from van der Drift and Boerrigter, 2006.)

of the FT process at the larger scales. The fossil fuel-based FT plants are huge, above 1000 MWth. With biomass, there may be a problem in supplying such a large process without extensive transport of biomass from distant sources. This may mean that any biomass-based plant is likely to be smaller at 100 MWth. However, there are ways to treat biomass to reduce its volume, so that it can be transported easily to the large central FT plant. The first is torrefaction, where biomass is heated at 250–300°C, which turns it into a brittle, solid mass that can be treated like coal. The second option is pyrolysis at 500°C that converts the biomass into oil/char slurry (Fig. 7.3).

At present syngas is mainly used by the chemical industry (Fig. 7.2), but some 8% (500 PJ per year) is used to produce fuels called GTL. FT processes are operated by Sasol in South Africa, and Shell in Bintulu, Malaysia. These are large plants of 1000 MWth due to the economies of scale and in one case use natural gas (CH_4).

To supply the EU-25, ten large plants of 1000 MWth would be required. At present, small- to medium-scale gasification systems of biomass are used for distributed heat and power (CHP) production. The larger scale of the GTL production also allows for the possibility for CO_2 capture and storage.

Bio-oil (Pyrolysis)

Fast pyrolysis of biomass in the absence of air at atmospheric pressure and 450–550°C will produce a mixture of gas, liquid and char (Fig. 4.7). The liquid is known as bio-oil, pyrolysis oil or bio-crude with yields as high as 80% depending on conditions. Bio-oil has a low calorific value at 16–18 MJ/kg but has the advantage of being a renewable fuel with low levels of sulfur and low net emissions of CO_2 (section 'Pyrolysis', Chapter 4). However, bio-oil is acidic, has a high viscosity, and is thermally unstable, and therefore requires processing before it can be used as a fuel. As a consequence bio-oil is regarded as a second-generation biofuel as it is not produced commercially at present. Bio-oil properties are compared with heavy fuel oil and diesel in Table 7.3, where the differences in viscosity and energy content are clear.

Table 7.3. Properties of bio-oil heavy fuel oil and diesel. (Adapted from Zhang *et al.*, 2007.)

Property	Bio-oil	Heavy fuel oil	Diesel
Moisture content (%)	15–30	0.1	
pH	2.5	–	
Density (kg/l)	1.2	0.94	0.84–0.85
Calorific value (MJ/kg)	16–19	40	38.5–45.6
Viscosity (cP)	40–100	180	2.8–3.51 cSt
Solids (%)	0.2–1.0	1	–

Bio-oil is a complex mixture containing some 300 compounds including acids, alcohols, aldehydes, esters, ketones, sugars, phenols, guaiacols, syringols, furans and lignin-derived compounds. Most of the compounds identified are phenols with aldehydes and ketones attached, which gives a high oxygen and highly hydrated content. The oxygen content needs to be reduced before the bio-oil can be used and the following methods have been used.

Hydrodeoxygenation

This is carried out with a hydrogen-providing solvent in the presence of catalysts (Co-Mo, Ni-Mo) at high temperatures and pressures.

Catalytic cracking

Oxygen can be removed from bio-oils by catalytic decomposition in the presence of catalysts. Although this is cheaper than hydrodeoxgenation, it suffers from high coking.

Steam reformation for hydrogen production

The production of hydrogen from the reforming of bio-oils has been investigated and shows some promise.

Emulsification

Bio-oils have been combined directly with diesel to form a fuel, but a surfactant is required as the bio-oil is immiscible with diesel. Chiaramonti *et al.* (2003) showed that the optimum level of bio-oil addition was between 0.5 and 2%, but above these values the viscosity was too high. Light fractions of bio-oil have been obtained by centrifugation and used at 10–30% in emulsions with diesel (Ikura *et al.*, 2003). The viscosity of the mixture was lower than the bio-oil and the cetane number was reduced by 0.4 for each 10% addition. In both cases, the long-term effect on the engine needs to be determined. The cost of bio-oil based on 2000 prices has been determined (Brammer *et al.*, 2006) at a value of €32/MWth which was not competitive with conventional energy sources.

Table 7.4. Applications of bio-oil. (Adapted from Brammer *et al.*, 2006.)

Application	Product
Boiler	Heat
Duel fuel IC diesel engine	Electricity
Duel fuel IC diesel engine	Combined heat and power (CHP)
Gas turbine	Electricity
Gas turbine	Combined heat and power (CHP)
Gas turbine combined cycle	Electricity
Gas turbine combined cycle	Combined heat and power (CHP)
Boiler, Rankine cycle	Combined heat and power
Diesel engine	Emulsion use for transport

Some of the applications of bio-oil as a heating fuel, diesel fuel and gas turbine fuel are listed in Table 7.4 (Brammer *et al.*, 2006). In six European countries, one application of bio-oil was competitive due to low biomass costs.

Hydrocarbons from Herbaceous Plants

A number of herbaceous plants accumulate long-chain hydrocarbons (terpenes) particularly those in the Euphorbiaceae such as the annuals *Hevea brasiliensis*, *Euphorbia lathyris* (3–10 t dry weight/ha/year) and *Calotropis procera* (10.8–21.9 t dry weight/ha/year). The hydrocarbons are produced, as latex, which consists largely of long-chain C 30 triterpenoids which can be cracked (pyrolysis) to form petrol and diesel. These herbaceous plants can be grown in various parts of the world and give quite good yields in terms of dry weights per hectare. Trees like *Eucalyptus globus*, *Pittosporum resiniferum* and *Copaifera multijuga* also produce oils, often in the fruit as in *P. resiniferum*, which can also be converted into petrol and diesel. However, with the Brazilian tree *C. multijuga*, the trunk can be tapped and the oil used directly as a diesel replacement.

Microalgal Hydrocarbons

A small number of algae are capable of producing terpenoid oils, one of which is *Botrycoccus braunii*, which is reported to accumulate up to 86% dry weight as oil (Dote *et al.*, 1993). Hydrocracking of the oil yielded 62% petroleum, 15% aviation fuel, 15% diesel and 3% heavy oil. The large-scale cultivation of this alga is under development.

Microalgal Oils

Microalgae can be used to produce a number of valuable products (Belarbi *et al.*, 2000; Del Campo *et al.*, 2000; Li *et al.*, 2001; Banerjee *et al.*, 2002), animal food (Knauer and Southgate, 1999), human health food (Becker, 2007) and as a wastewater treatment (Travieso *et al.*, 2002; Kebede-Westhead *et al.*, 2006). In addition to these options, microalgae have been proposed as systems for the sequestration of CO_2

(Sawayama *et al.*, 1995; Zeiler *et al.*, 1995; de Morais and Costa, 2007a,b) and the production of biofuels (Chisti, 2007). The biofuels include biogas (CH_4) by anaerobic digestion of the biomass (Spolaore *et al.*, 2006), biodiesel from microalgal oils (Nagle and Lemke, 1990; Sawayama *et al.*, 1995; Minowa *et al.*, 1995; Miao and Wu, 2006; Xu *et al.*, 2006; Chisti, 2007), hydrogen (Fedorov *et al.*, 2005) and the direct use of algae in emulsion fuels (Scragg *et al.*, 2003).

Biodiesel is one of the sustainable fuels which can replace diesel as a transport fuel and is usually made by the transesterification of plant-derived oils, waste cooking oils and animal fats. However, microalgae should be considered as another source of biodiesel because of the following:

- They have higher photosynthetic efficiency than terrestrial plants.
- They have rapid growth rate, with doubling times of 8–24 h.
- They have high lipid content of 20–70%.
- They facilitate direct capture of CO_2, 100 t algae fix ~183 t CO_2.
- They can be grown on a large scale.
- They will not compete with terrestrial plants in food production.
- They produce valuable products.
- They include freshwater and marine species.
- They have a much better yield of oil per hectare, oil palm 5000 t/ha, algae 58,700 t/ha (Table 7.5; Chisti, 2007).

The use of microalgal oil to produce biodiesel is very much in the developmental stage, and so it should be regarded as a third-generation biofuel.

To use microalgae for the production of biodiesel a number of processes must be carried out and these are outlined in Fig. 7.7 and consist of strain selection, large-scale cultivation, harvesting, extraction of the oil, production of biodiesel from the oil, and the economics of the process.

Strain selection

Not all microalgae accumulate high concentration of oil, but there are a number of freshwater and marine species that do. Some examples of the oil levels accumulated

Table 7.5. Comparison of an open raceway and closed photobioreactor. (Adapted from Chisti, 2007.)

Parameter	Raceway	Photobioreactor
Biomass production per year (kg)	100,000	100,000
Volumetric productivity (kg/m³/day)	0.117	1.535
Areal productivity (kg/m²/day)	0.035	0.072
Biomass concentration (kg/m³)	0.14	4.00
Dilution rate (day⁻¹)	0.25	0.384
Area needed (m²)	7,828	5,681
Oil yield (m³/ha)[a]	42.6 (37.5)[b]	58.7 (51.6)
Carbon dioxide consumption per year (kg)	183,333	183,333

[a]Based on 30% oil in biomass.
[b]( ) values in tonnes per hectare; compare this with rapeseed at ~1 t/ha.

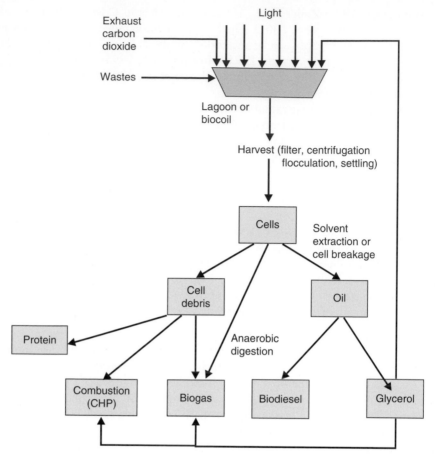

Fig. 7.7. Various ways of utilizing microalgae for the production of energy including biodiesel.

are given in Chisti (2007) and Scragg (2005). However, not all microalgal oils are suitable for biodiesel production, as some contain high levels of unsaturated fatty acids which reduce the oxidative stability of the biodiesel. In many cases, high oil accumulation is only found under some form of stress such as nitrogen limitation (Illman *et al.*, 2002), and so growth may have to be in two stages in order to obtain high levels of oil. In contrast, heterotrophic growth in glucose stimulated oil accumulation in *Chlorella prototothecoides* (Miao and Wu, 2006; Xu *et al.*, 2006). Strain selection will also be important depending on the type of cultivation system used.

Large-scale cultivation

There is a considerable body of information on the large-scale cultivation of microalgae in bioreactors of various designs (Molina Grima *et al.*, 2001; Scragg *et al.*, 2002; Acien Fernandez *et al.*, 2003; Chisti, 2007; de Morais and Costa, 2007a,b). The designs of photobioreactors can be divided into two types – open and closed – and the advantages and disadvantages of these types are outlined below:

Open bioreactors:

- Natural water, raceway ponds, inclined surfaces. These can suffer from: water and CO_2 loss, contamination and pollution, requirement of large area, limitation on the number of species that can be grown, no process control, dependency on weather, poor mixing and low biomass (0.1–0.2 g/l).

Closed bioreactors:

- Stirred vessels, tubular bioreactor, laminar bioreactor, plastic bag vessels. Best for high-value products, process control, continuous culture possible, all types of algae grown, flexible production, not affected by weather, high biomass (2–8 g/l).

It would appear that the two best bioreactor designs are the raceway and tubular designs. The raceway is considerably simpler but mixing is limited, temperatures vary and the biomass concentration is low. The tubular bioreactors are enclosed and with good mixing and circulation a high biomass concentration can be achieved without contamination. They are, however, more expensive to operate and require cooling during daylight. A wide range of bioreactor designs have been used to culture micro-algae and some examples of the various designs are given in Table 7.6.

Table 7.6. Alternative photobioreactor designs.

Bioreactor design	Algal species	Reference
Closed		
Vertical tube	*Chlorella* sp.	de Morais and Costa (2007b)
	Scendesmus obliquus	
	Spirulina sp.	
Parabola, pipe, diamond	*Chaetoceros calcitrans*	Sato *et al.* (2006)
Tubular horizontal	*Spirulina platensis*	Richmond *et al.* (1993); Acien Fernandez *et al.* (2003); Molina Grima *et al.* (2001)
Horizontal tubular	*Phaeodactylum tricornutum*	Miron *et al.* (1999)
L-shaped	*Euglena gracilis*	Chae *et al.* (2006)
Internally illuminated stirred tank	*Chlorella pyrenoidosa*	Ogbonna *et al.* (1996)
5 l stirred tank bioreactor	*Isochrysis galbana*	Molina Grima *et al.* (1993)
Plexiglas annular	*Nannochloropsis* sp.	Zittelli *et al.* (2003)
Tubular/flat	*Spirulina* sp.	Tredici and Zittelli (1998)
Dual sparging column	*Rhodamonas* sp.	Eriksen *et al.* (1998)
Cone-shaped helical tubular	*S. platensis*	Watanabe and Hall (1996)
Helical tubular	*Chlorella* sp.	Scragg *et al.* (2003)
Vertical flat plate	*Synechocystis aquatilis*	Zhang *et al.* (2002)
Light supplied by optical fibres	*Porphyridium purpureum*	Fleck-Schnieder *et al.* (2007)
Stirred draft tube		
Open		
Inclined	*Chlorella* sp.	Doucha and Livansky (2006)
Open tanks 220 l	*Chaetoceros* sp.	Csordas and Wang (2004)
0.7 m deep polyethylene tanks	*Chaetoceros* sp.	Elias *et al.* (2003)
Raceways	*Spirulina* sp.	Olquin *et al.* (1997)

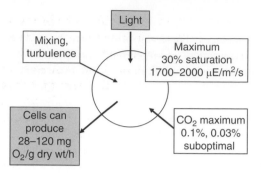

Fig. 7.8. Outline of the parameters which affect the growth of algae in bioreactors.

Whatever design is used, microalgal growth in bioreactors is influenced by four parameters: the supply of light, the supply of CO_2, mixing (turbulence) and the build-up of oxygen (Grobbelaar, 1994) (Fig. 7.8). The maximum amount of light is about 30% saturation, and values above this can cause photobleaching, the loss of chlorophyll. The CO_2 levels of 0.03% are below the optimum for growth, concentration of 0.1% is more suitable, and there have been cases where 10% CO_2 did not inhibit growth. In the light, microalgae can produce oxygen rapidly and a build-up of oxygen can inhibit growth. Mixing in the form of turbulence is essential to keep the cells in suspension and for gaseous exchange.

Commercial production of microalgae has been used to produce pigments, food supplements and shellfish food. The designs that have been used to produce microalgae commercially are given in Table 7.7.

Harvesting

The two critical stages in product development are the harvesting of the microorganisms and the extraction of the product as these unit operations can add considerable costs to the process. For microalgae, there are a number of methods for harvesting

Table 7.7. Commercial microalgae cultivation. (Adapted from Borowitzka, 1999.)

System	Algae	Max. volume	Location
Tanks	Many species	1×10^4	Worldwide
Extensive open ponds	Dunaliella salina	1×10^9	Australia
Circular ponds with rotating arm	Chlorella sp.	1.5×10^4	Taiwan, Japan
Raceways	Chlorella sp., Spirulina sp., D. salina	3×10^4	Japan, Taiwan, USA, Thailand, China, India, Vietman, Chile, Israel
Large bags	Many species	1×10^3	Worldwide
Bioreactors heterotrophic	Chlorella sp., Crypthecodinum cohnii	$>10^3$	Japan, Taiwan, Indonesia, USA
Two-stage (indoors and then outdoors in a paddlewheel pond)	Haematococcus pluvialis	–	USA

the cells including centrifugation, filtration, flocculation and settling. Flocculation can be used to improve the other methods of harvesting. Flocculation can be carried out using multivalent metal salts (Molina Grima *et al.*, 2003) or cationic polymers. Centrifugal recovery is rapid and expensive but has been used for many microalgae. Filtration using filter presses has proved unsuccessful with the smaller microalgae, and membrane filtration has not been extensively used.

Extraction

Intracellular oils are difficult to extract from wet biomass (Belarbi *et al.*, 2000), but can be more easily extracted from freeze-dried cells or cell paste. Oil has been extracted from *Phaeodactylum tricormutum* (diatom) and *Monodus subterraneus* (green alga) with solvents under pressure (Belarbi *et al.*, 2000). In the case of *C. protothecoides*, the cells were freeze-dried before solvent extraction as were the oils from *Isochrysis galbana* (Molina Grima *et al.*, 1994). Microalgal cells may be disrupted to extract the oils using a number of microbial cell disruption methods, but these methods can be expensive. Free fatty acids have been extracted from wet biomass using a potassium hydroxide-ethanol mixture (Molina Grima *et al.*, 2003). Whole cells of *Dunaliella tertiolecta* have been liquefied at 300°C and 10 MPa to form oil comparable to fuel oil. Both supercritical CO_2 (Mendes *et al.*, 2003; Gouveia *et al.*, 2007) and thermochemical liquefaction have also been used to produce biodiesel from macroalgae (Aresta *et al.*, 2005). In addition, whole microalgal cells containing high levels of oil have been used directly in diesel and biodiesel in emulsion fuels (Scragg *et al.*, 2003). In general, all methods both mechanical and solvent based are expensive and will affect the cost of the biofuel.

Production of microalgal biodiesel

Microalgal biodiesel will need to comply with the standard EN 14214 in the EU and ASTM D 6751 in the USA before it can be universally accepted. Microalgal oils tend to contain more polyunsaturated fatty acids than plant oils, and those with four or five double bonds are more susceptible to oxidative degradation. However, the biodiesel produced from oil extracted from *C. protothecoides* had characteristics which were similar to diesel (Miao and Wu, 2006) apart from a slightly higher viscosity (Table 7.8).

Table 7.8. Comparison of the properties of microalgal biodiesel and diesel. (From Miao and Wu, 2006; Xu *et al.*, 2006; Stanhope-Seta, 2007.)

Properties	Microalgal biodiesel	Diesel	En 14214 specifications
Density (kg/l)	0.864	0.838	0.86–0.90
Viscosity (mm² s⁻¹ cSt at 40°C)	5.2	1.9–4.1	3.5–5.0
Flash point (°C)	115	75	>101
Pour point (°C)	−12	−50–10	–
Cold filter plugging point (°C)	−11	−3–6.7	Summer – 0 Winter −15
Acid value (mg KOH/g)	0.374	Max 0.5	0.5
Heating value (MJ/kg)	41	40–45	–

Biodiesel from Plant Oils and Animal Fats (Esters)

Untreated plant-derived oils have been used to replace diesel in emergency situations and Diesel's patent included plant-derived oils as one of the fuels. However, long-term use of untreated oil does cause problems with diesel engines. Recently interest has been revived in the use of oils as a renewable and carbon-neutral replacement for diesel.

Oils are produced from plants throughout the world in considerable quantities (Shay, 1993). Plant oils are normally extracted from oil-containing seeds, where the plant uses oil rather than starch as an energy store for the seed. Seed oil can be extracted from a wide range of annual crops such as soybean, sunflower, rapeseed (canola) and the perennial oil palm. A list of high oil-producing plants is given in Table 7.9 where it is clear that perennial crops have a higher yield of oil per hectare. Despite the higher yields from the perennial plants, annual crops like rapeseed and soybean have commanded most interest, probably because there is already a market for their oil and annuals are a more flexible crop which are often grown in rotation.

The advantages of using plant-derived oils are:

- They are liquid.
- Their calorific content is 80% of diesel.
- They are readily available in large quantities disregarding competition with food crops.
- They are renewable/sustainable as derived from crops.
- They are non-toxic and much more biodegradable than diesel.
- They are CO_2-neutral, combustion releases CO_2 previously fixed by the plant.
- They are contain no sulfur.

However, the following are problems associated with the use of untreated plant oils:

- They have high viscosity.
- They have low volatility and high flash point.

Table 7.9. Oil yields from annual and perennial crops. (From Shay, 1993.)

Plant	Yield (kg/ha/year)
Annuals	
Cotton (*Gossypium hirsutum*)	273
Soybean (*Glycine max*)	375
False flax (*Camelina sativus*)	490
Mustard seed	480–1000
Safflower (*Carthamus tinctorius*)	655–1040
Sunflower (*Helianthus annuus*)	800
Rapeseed (*Brassica napus*)	1000
Castor bean (*Ricinus communis*)	1188
Jojoba (*Simmondsia chinensis*)	1528
Perennials	
Jatropha curcas	759–1590
Olive (*Olea europea*)	1019
Coconut (*Cocos nucifera*)	2260
Oil palm (*Elaeis quineensis*)	5000

Table 7.10. A comparison of the properties of diesel and plant oil.

Property	Diesel	Rapeseed oil
Density (kg/l)	0.84	0.778–0.91
Viscosity (cSt)	2.8–3.5	37–47
Flash point (°C)	64–80	246–273
Cetane number[a]	48–51	38–50
Calorific value (MJ/kg)	38.5–45.6	36.9–40.2

[a]Cetane number is an indicator of the ignition quality of the fuel and is linked to ignition delay. Standards have been set for cetane number measured against hexadecane (cetane) assigned a value of 100.

- They contain reactive unsaturated hydrocarbon chains.
- They have carbon deposits.

To function correctly in a diesel engine, the fuel must form a fine mist, which should burn rapidly and evenly. Untreated plant-derived oil contains residual components such as waxes, gums and high molecular weight fatty components, which clog the fuel lines and filters. High oil viscosity cause poor atomization, affecting ignition and combustion, which gives carbon deposits on injectors, combustion chamber walls and pistons. The polymerization of unsaturated fatty acids in the combustion chamber also causes deposits on the wall, and some components mix with the lubricating oils increasing their viscosity (Peterson *et al.*, 1996; Ma and Hanna, 1999). The presence of water in the oils can allow microbial growth that can block the fuel filters. To illustrate the problem of viscosity, a comparison of the properties of diesel and plant oil is shown in Table 7.10.

Different methods have been used to reduce the viscosity of the oil, which includes blending with diesel, microemulsification, pyrolysis and transesterification. Of these methods, only transesterification has been successful, and the mixture of esters formed is called biodiesel.

Transesterification

Biodiesel is a replacement for diesel and is produced by reacting plant oils and animal fats with an alcohol to form a mixture of fatty acid esters in a reaction known as transesterification. Biodiesel is available commercially and should be regarded as a first-generation biofuel. The idea of splitting the triglycerides in fats and oils and using the resulting esters as a fuel has been around for a considerable time. Walton, in 1938, suggested the splitting of triglycerides (Graboski and McCormick, 1998), and there is a report of fatty acid esters being used as a fuel in the Congo in 1937 (Knothe, 2001). Subsequently, there have been a number of reports of using plant oil/diesel blends in engines where the problems of high viscosity of oil were encountered. One of the first reports of the use of esters was in 1980 using sunflower oil esters which appeared to remove many of the problems associated with untreated oils, in particular, viscosity. Since then, there has been a considerable number of reports on the production of fatty acid esters from a wide range of fats and oils. The European quality standards for fatty acid methyl esters, known as biodiesel, came into force in 2004 and are known as EN 14214 (biodiesel) and EN 14213 (heating fuel) (Schober *et al.*, 2006).

Transesterification of plant oils is the conversion of the triglycerides which make up oils into fatty acid esters and glycerol. Triglycerides are the main component of fats and oils and consist of three long-chain fatty acids linked to a glycerol backbone. When the triglyceride reacts with an alcohol, the three fatty acids are released and combined with the alcohol to form alkyl esters. Transesterification of pure oils can be carried out rapidly with methanol and NaOH as the catalyst (Van Gerpen, 2005). Methanol is normally used as the alcohol, although ethanol, 2-propyl and 1-butyl will also suffice (Lang *et al.*, 2001).

$$
\begin{array}{ccccc}
CH_2 & & & CH_2OH & R^1COOCH_3 \\
| & & \overset{\text{NaOH catalyst}}{=} & | & \\
CH_2 & +3\ CH_3OH & & CHOH & + \quad R^2COOCH_3 \qquad (7.4) \\
| & & & | & \\
CH_2 & & & CH_2OH & R^3COOCH_3 \\
\text{triglyceride} & \text{methanol} & & \text{glycerol} & \text{methyl esters}
\end{array}
$$

The reaction can be catalysed by alkalis, acids, lipase enzymes and inorganic heterogeneous catalysts (Fukuda *et al.*, 2001; Vincente *et al.*, 2004). The conditions for catalysis are a temperature near to the boiling point of methanol (60°C), although room temperature will suffice with pure oil, a molar ratio of alcohol/oil of between 3:1 and 6:1, and NaOH as the catalyst. The stoichiometric molar ratio of methanol/oil is 3:1 but in order to drive the reaction towards ester formation the ratio is increased to ratios of up to 9:1. The effect of the molar ratio of methanol/oil on the process of transesterification is shown in Fig. 7.9.

The transesterification reaction requires catalysis and apart from alkali catalysts others have been used including acids, enzymes and solid catalysts (Suppes *et al.*, 2004; Vincente *et al.*, 2004; Meher *et al.*, 2006a). The alkali-catalysed transesterification is by far the fastest process (Fig. 7.10), but is sensitive to impurities in the raw materials.

The presence of water and free fatty acids in the oil consumes alkali, and forms soaps which in turn produce emulsions. Emulsions stop the separation of glycerol as the reaction proceeds, which reduces the yield of biodiesel (Fig. 7.11).

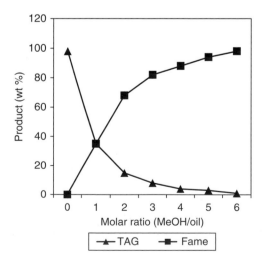

Fig. 7.9. The effect of the methanol/oil ratio on methyl ester production. MeOH, methanol; TAG, triacylglycerols; FAME, fatty acid methyl esters. (Redrawn from Freedman *et al.*, 1986.)

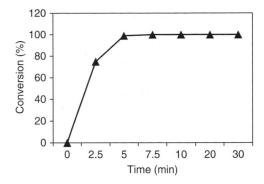

Fig. 7.10. The production of methyl esters during NaOH-catalysed transesterification. (Redrawn from Freedman *et al.*, 1986.)

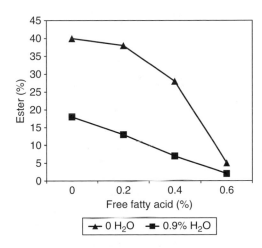

Fig. 7.11. Effect of the presence of free fatty acids and water on the NaOH-catalysed transesterification of beef tallow. (Redrawn from Ma *et al.*, 1998.)

$$R\text{-COOH} + KOH = R\text{-COO}^-K^+ + H_2O \qquad\qquad (7.5)$$
$$\underset{\text{fatty acid}}{} \qquad\qquad \underset{\text{potassium soap}}{}$$

In extreme cases, the treated oil will set into a gel formed from a combination of glycerol and soap. An ester yield of less than 5% was obtained in the presence of 0.6% free fatty acids (Canakci and van Gerpen, 1999; Usta, 2005). Therefore, oils containing no water and less than 0.5% free fatty acids are required for successful alkali catalysis. These properties can be obtained with most plant oils, but waste cooking oils, rendered fats and some plant oils contain between 0.7 and 24% water and 0.01–75% free fatty acids (Zhang *et al.*, 2003; Meher *et al.*, 2006a; Canakci, 2007). Unfortunately, there are large amounts of unrefined plant oils, waste cooking oils and soapstocks available for biodiesel production. Acid catalysts, mainly sulfuric, hydrochloric and phosphoric acids, have not been used widely as the reaction is very much slower than the alkali catalysts (Fig. 7.12), but acid catalysis is not affected by free fatty acids.

Therefore, a two-stage process has been developed where in the first stage acid catalysis is used to esterify the free fatty acids, and the alkali-catalysed system is used in the second stage to transesterify the triglycerides (Zullaikah *et al.*, 2005; Wang *et al.*, 2006) (Fig. 7.13).

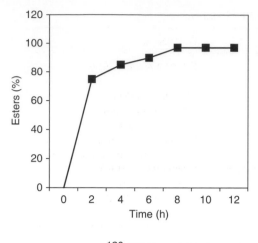

Fig. 7.12. Acid-catalysed esterification of rice bran oil containing 75.8% free fatty acids. (From Zullaikah *et al.*, 2005.)

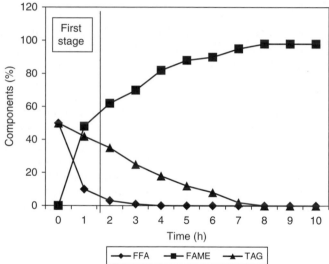

Fig. 7.13. The two-stage production of biodiesel from oil containing 50% free fatty acids. Stage one is catalysed by sulfuric acid and the second is alkali-catalysed. FFA, free fatty acids; FAME, fatty acid methyl esters; TAG, triacylglycerols. (From Zullaikah *et al.*, 2005.)

Alternative catalysts

However, alkalis and acids are not the only catalysts which can be used in the transesterification reaction and these include enzymes and solid catalysts. Some of the solid catalysts are listed in Table 7.11.

Transesterification using heterogeneous catalysts has been investigated using basic zeolites and alkaline metal compounds. Metal oxides, hydroxides and alkoxides have been used to transesterify rapeseed oil (Gryglewicz, 1999) where calcium oxide was the most effective. Metal oxides and those loaded with Al_2O_3, SiO_2 and MgO were also used to treat rapeseed oil (Peterson and Scarrach, 1984).

Oil extracted from *Pongamia pinnata* has been transesterified using a solid Li/ CaO catalyst even in the presence of 0.48–5.75% free fatty acids (Meher *et al.*,

Table 7.11. Solid catalysts used to produce biodiesel.

Oil	Catalyst	Reference
Soybean	Zeolite	Suppes *et al.* (2004)
	Metals (Ti, Si)	
Jatropha curcas	Calcium oxide	Zhu *et al.* (2006)
Pongamia pinnata	Calcium oxide	Meher *et al.* (2006b)
Glyceryl tributyrate	Li-calcium oxide	Watkins *et al.* (2004)
Soybean	Lewis acid	Di Serio *et al.* (2005)
Rapeseed	Metal oxides	Peterson and Scarrach (1984)
Rapeseed	Metal oxides, hydroxides,	Gryglewicz (1999)
Mixture of oils	Fe-Zn cyanide complex	Sreeparasanth *et al.* (2006)
Soybean oil	Solid super acid	Furuta *et al.* (2004)
	(sulfated Zi and Sn)	

2006b) and *Jatropha curcas* oil using CaO (Zhu *et al.*, 2006). A number of modified zeolites have been used successfully to transesterify soybean oil (Suppes *et al.*, 2004). Much of the research has been with solid base catalysts but solid acid catalysts have also been used. Tungstated zirconia, a solid super acid catalyst, has been used to transesterify soybean oil at 200–300°C, and has given a conversion of over 90% (Furuta *et al.*, 2004). More recently, amorphous zirconia combined with titanium and aluminium has been shown to give over 95% conversion of soybean oil at 250°C (Furuta *et al.*, 2006).

Microbial lipases have the ability to transesterify oils in the presence of methanol. These enzymes function in the presence of water and the catalyst and salts do not need removing at the end of the reaction (Table 7.12). However, the enzymes are more expensive than the simple inorganic catalysts. Some of the expense of using enzymes can be reduced by enzyme immobilization which allows a continuous process and increases the working life of the enzyme (Ban *et al.*, 2001; Fukuda *et al.*, 2001).

Table 7.12. Enzymatic transesterification.

Oil	Lipase	Conversion (%)	Reference
Rapeseed	*Alcaligenes* sp. immobilized on activated bleaching earth	80	Du *et al.* (2006)
Rapeseed	*Candida rugosa*	97	Linko *et al.* (1998)
Sunflower	*Mucor meihei*	83	Selmi and Thomas (1998)
Waste cooking grease	*Pseudomonas cepacia* and *Candida antarctica*	85.4	Wu *et al.* (1999)
Sunflower	*Pseudomonas fluorescens*	82	Mittelbach (1990)
Palm kernel	*P. cepacia*	15–72	Abigor *et al.* (2000)
Soybean	*Rhizopus oryzae* immobilized	90	Ban *et al.* (2001)
Cotton seed oil	*C. antartica*	100	Royon *et al.* (2007)
Soybean oil	*C. antartica*	>90	Watanabe *et al.* (2002)

Transesterification has also been carried out using supercritical methanol, ethanol, propanol and butanol. The process does not require a catalyst but high temperatures (~300°C) and pressures (8 MPa) (Cao *et al.*, 2005; Demirbas, 2006a,b).

Properties of Biodiesel

Biodiesel is defined as a mixture of fatty acid esters, normally methyl esters, produced from plant oils, animal fats and waste cooking oils. The mixture of fatty acid esters (biodiesel) has properties very similar to those of diesel and can be used with very little modification in a diesel engine. A comparison between conventional diesel, plant oils and biodiesel is shown in Table 7.13.

As can be seen, the main differences between diesel and the widely used plant oil from rapeseed are higher viscosity and higher flash point. The elevated flash point is an advantage as it makes the oil safer, but the higher viscosity makes the long-term use of the oil in engine difficult, especially in cold weather. However, by forming a fatty acid ester mixture, the viscosity is considerably reduced to a level where the biodiesel can be used in diesel engines without modification. The European standard for 100% biodiesel is EN 14214 and the values are given in Table 7.13.

Effect of Fatty Acid Ester Content on the Properties of Biodiesel

The fatty acid profile of biodiesel will reflect the fatty acid profile of the oil used in its production. Table 7.14 gives the fatty acid profiles of biodiesel produced from a variety of oils. The most striking differences are perhaps the biodiesel produced from palm, coconut and linseed oil. Both palm and coconut oil contain high levels of lauric acid and linseed a high concentration of linolenic acid. This will have a profound effect on the characteristics of the biodiesel. The high level of saturated fatty acids in palm biodiesel will mean poor cold temperature characteristics with a cloud point of 8°C and a pour point of 6°C. Linseed biodiesel has a high proportion of unsaturated fatty acid esters and should exhibit rapid oxidation and polymer formation. The properties of biodiesel is affected by the proportion of long- and short-chain fatty acids and the presence of one or more double bonds (saturated and unsaturated) (Fig. 7.14).

Table 7.13. A comparison of the properties of diesel and biodiesel.

Property	Diesel	Rapeseed oil	Rapeseed methyl esters (biodiesel)	Biodiesel specification EN 14214
Density (kg/l)	0.84–0.85	0.778–0.91	0.768–0.88	0.86–0.9
Viscosity (cSt)	2.8–3.51	37–47	6.1–7.2	3.5–5.0
Flash point (°C)	64–80	246–273	170–185	>101
Cetane number	47.8–51	37.6–50	51.8–54.4	>51
Calorific value (MJ/kg)	38.5–45.6	36.9–40.2	35.3–40.5	Na
Cloud point (°C)	−18	−3.9	−4 to −1	Na
Pour point (°C)	−26	31.7	−7 to −9	Na

Table 7.14. Fatty acid profiles of methyl esters in biodiesel. (From Graboski and McCormick, 1998; Lang *et al.*, 2001; Mittlebach and Gangl, 2001; Haas, 2005; Hu *et al.*, 2005; Schober *et al.*, 2006.)

Oil	Lauric C14	Palmitic C16	Stearic C18	Oleic C18:1	Linoleic C18:2	Linolenic C18:3	Erucic C22:1
Rapeseed		4.58	2.0	60.0	21.66	7.92	0.79
		4.8	1.8	62.2	19.9	8.9	0.2
		4.2	2.2	67.2	18.9	7.4	0
Soybean		10–12	4.0	22–25	53	6–7.9	–
Sunflower		5.8	5.7	20.4	66	–	–
Maize		12.3	2.0	29.8	54.7	0.5	–
Palm	38.2	5.98	18.5	10.2	3.2	0.14	–
Soybean soapstock		16.4	4.8	16.5	55.3	7.0	–
Coconut	37.8	7.2	18.7	12.3	4.5	0.16	–
Linseed	–	5.2	3.2	14.5	15.3	61.9	–

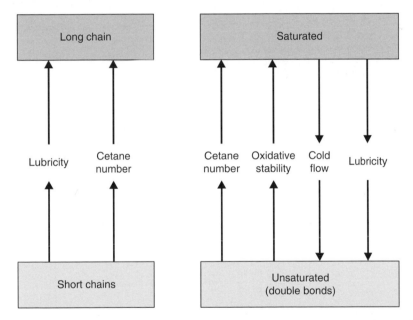

Fig. 7.14. The effect of chain length and degree of saturation on cetane number, oxidative stability, lubricity and cold flow.

Biodiesel has a similar energy content and viscosity as diesel in addition to a number of positive attributes such as increased flash point, non-toxic, rapidly biodegradable, lower emissions and increased lubricity (Table 7.15). Negative aspects are the poor low-temperature characteristics and oxidative stability. Many of these properties are affected by the fatty acid ester content depending on chain length and the presence of double bonds.

Other parameters which are included in the table are acid value and oxidative stability. Acid value is the measure of the unsaturation of the fatty acids in the mixture

Table 7.15. Properties of various biodiesel. (From Graboski and McCormick, 1998; Sheehan et al., 1998; Williamson and Badr, 1998; Srinvastava and Prasad, 2000; Mittlebach and Gangl, 2001; Al-Widyan et al., 2002; Antolin et al., 2002; Haas, 2005; Sarin et al., 2007.)

Oil source	Density (mg/ml)	Cetane number	Energy content (kJ/kg)	Viscosity (cPs)	Cloud point (°C)	Pour point (°C)	Flash point (°C)	Acid value (gI$_2$/100g)	Oxidation stability (h)
Soybean	0.88	52.5	40.0	4.2	2	−1	169	0.15	3.8
Rapeseed	0.88	52.2	40.5	5.8	−1	−6	162	0.16	5.6
								0.32	
Sunflower	0.88	48.0	40.0	4.22	0	−4	183	0.179	1.73
								0.2	
Cotton	0.88	51.2	40.2	–	–	3	110	–	–
Palm oil	0.87	53.0	39.5	4.5	8	6	174	0.24	13.37
Tallow	0.88	58.8	40.0	4.45	12	9	106		–
Linseed	0.88	–	40	3.32	0	–	–	0.335	–
Jatropha	–	57.1	–	4.4	4	–	163	0.48	3.23
Pongamia	–	55.1	–	4.16	4	–	141	0.1	2.35
Cooking oil	0.88	–	39.3	15.1	0	0	109		
Soybean soapstock	0.88	51.3	–	4.3	6	–	169		

measured as $gI_2/100\,g$ sample. Oxidative stability is the time required to induce the production of volatile breakdown products when incubated at elevated temperatures, in hours.

Cetane number

Cetane numbers rate the ignition properties of diesel fuel as a measure of the fuel's ignition on compression as measured by ignition delay. Cetane affects smoke production on start-up, drivability before warm-up and diesel knock at idle. Cetane number is measured in a single-cylinder engine compared with reference blends of n-cetane and heptane. Cetane numbers for biodiesel are influenced by the fatty acid ester profile making up the biodiesel (Table 7.15). In general, as the number of carbon atoms in the fatty acid esters increases, so does the cetane number (Fig. 7.15). However, as the number of double bonds increases the cetane number decreases (Fig. 7.16), so for a high cetane number long-chain saturated fatty acids are needed (Graboski and McCormick, 1998; Knothe *et al.*, 2003). The formation of pollutants is dependant on ignition delay and cetane number and a high cetane number gives less NO_x.

Cold flow

One of the major problems with biodiesel is its poor low-temperature flow properties shown by its high cloud and pour points. At the cloud point, long-chain fatty

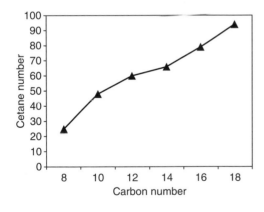

Fig. 7.15. The effect of ester chain length on cetane number. (Redrawn from Graboski and McCormick, 1998.)

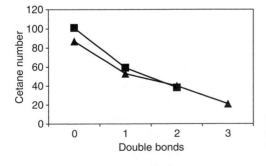

Fig. 7.16. The effect of double bonds on cetane number. (From (▲) Graboski and McCormick, 1998; (■) Knothe *et al.*, 2003.)

acid esters begin to form small wax crystals and when these reach 5 μm in size, the fuel begins to look cloudy. As the temperature decreases, the crystals grow and aggregate in a form which can plug filters and eventually it ceases to pour. Often low-temperature filterability (LTFT) is used as a low-temperature characteristic which is generally halfway between cloud and pour point. Several methods have been used to improve low-temperature characteristics including additives, branched chain esters and winterization.

Treatment with chemical additives, pour depressants, is one way that conventional diesel is modified to improve its low-temperature characteristics.

The molecular structure of pour depressants is polymeric hydrocarbon chains with polar groups which function by adsorption, co-crystallization, nucleation and improved wax solubility. Vegetable oils lack polar groups but ozonation can reduce the pour point (Soriano *et al.*, 2005). Other cold flow improvers have been tested, and these reduced the pour point but had little effect on cloud points (Chiu *et al.*, 2004).

Using branched alcohols, such as iso-propanol or iso-butyl alcohol in place of methanol and ethanol in the transesterification process, yields esters which have lower pour points. However, this appeared to increase the viscosity of the biodiesel and also resulted in incomplete esterification.

Winterization of biodiesel has been carried out which increases the unsaturated fatty acid esters. The saturated fatty acids are removed as they have higher melting points. The process reduces biodiesel yields and the high concentration of saturated fatty acids means a lower cetane value.

Oxidative stability

Biodiesel is subject to oxidative breakdown which is related to the double bond content of the fatty acid methyl esters, and oxidation can lead to increased acidity, formation of shorter fatty acids and the production of gums. This is an important feature, as stability is needed if biodiesel is to survive long-term storage, particularly for biodiesel from waste oils that have lost their natural antioxidants.

During oxidation the fatty acid methyl ester form a radical next to the double bond which binds oxygen forming a peroxide radical. The peroxide radical reacts with a fatty acid forming an acid releasing the radical and thus forming an autocatalytic cycle.

$$RH + I \rightarrow R^* + IH \tag{7.6}$$

$$R^* + O_2 \rightarrow ROO^* \tag{7.7}$$

$$ROO^* + RH \rightarrow ROOH + R^* \tag{7.8}$$

The carbons most susceptible to free radical formation are those adjacent to the double bonds. The fatty acid with OOH group decomposes into aldehydes, hexanals, heptanals, propanol, aliphatic alcohols, formic acid and formate esters. Shorter fatty acids can be formed and some can link together to form polymers (gums). Therefore, saturated fatty acid esters will be more resistant to oxidation.

Oxidative stability is measured using the Rancimat, where the sample is heated for 6 h at 110°C and air is passed through the sample and collected in a separate vessel where the conductivity is measured (Fig. 7.17) (Sarin *et al.*, 2007).

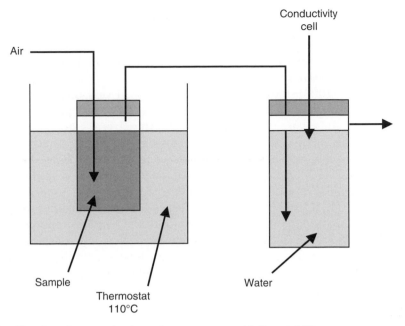

Fig. 7.17. The Rancimat method used to measure oxidative stability.

The first study on the oxidative stability of biodiesel was by du Plessis *et al.* (1985) using sunflower-derived biodiesel stored at different temperature for 90 days. Other studies by Bondioli *et al.* (2004) and Mittelbach and Gangl (2001) on the storage of rapeseed-derived biodiesel were run over 1 year and 200 days, respectively. In the study by Mittelbach and Gangl (2001), rapeseed biodiesel was stored in polyethylene bottles at 20–22°C open or closed, in the light and in the dark. Samples were removed at intervals and the oxidative stability measured using the Rancimat system (Fig. 7.18). In the Rancimat system, the sample is heated to 100°C and air is passed

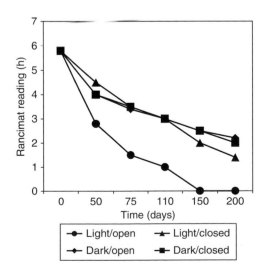

Fig. 7.18. The effect of storage conditions on the stability of biodiesel. (Redrawn from Mittelbach and Gangl, 2001.)

through the liquid. The air is run into a conductivity cell filled with water. After a few hours the conductivity in the cell increases rapidly as a result of volatile organic acid compounds produced by the oxidative breakdown of the sample collecting in the cell. It is the time required for the induction of the increase in conductivity that is taken as a measure of oxidative stability.

Bondioli *et al.* (2003) studied the long-term storage, 1 year, of 11 different biodiesel preparations, some containing antioxidants. The list of samples and the results after 12 months' storage are given in Table 7.16. After 1 year's storage, a number of the parameters measured did not change but oxidative stability as revealed by the Rancimat data did change. Those samples with the lowest value changed less, such as the distilled sunflower biodiesel. The Rancimat values decreased with time and were dependent on both the state of the sample and the storage conditions. The addition of two antioxidants TBHQ and pyrogallol had little effect on stability in this case.

To combat the oxidation of biodiesel, natural and synthetic antioxidants have been added. Crude and distilled palm oil methyl esters were tested for their oxidative stability and these were 25.7 and 3.52h in the Rancimat system (Liang *et al.*, 2006). The crude palm oil methyl ester mixture contained 644 ppm vitamin E (α-tocopherol) and 711 ppm β-carotene, whereas the distilled version contained very little of these compounds. Various quantities of three antioxidants α-tocopherol, butylated hydroxytoluene (BHT) and tert-butyl hydroquinone (TBHQ) were added to the distilled methyl ester mixture. The two synthetic antioxidants were better than the natural antioxidants and 50 ppm was sufficient to achieve the EN 14214 Rancimat standard. Dunn (2005) has also determined the effect of different antioxidants on soybean biodiesel and the NREL report indicates that antioxidants will be required for extended storage.

Another approach has been to blend Jatropha and palm oil biodiesel. Palm oil biodiesel contains high levels of palmitic (C16:0) and oleic (C18:1) acids which are

Table 7.16. The storage stability of a number of biodiesel samples after 12 months storage. (Adapted from Bondioli *et al.*, 2003.)

Sample	Peroxide value (meqO$_2$/kg)		Viscosity (mm^2/s)		Rancimat induction time (h)	
	Before	After	Before	After	Before	After
Rape	7.3	11.4	4.37	4.49	7.51	6.20
Rape + TBHQ 400 mg/kg	2.3	5.4	4.41	4.50	36.0	32.77
Rape: low stability	10.2	20.5	4.36	4.52	6.3	1.24
Rape	3.4	13.3	4.41	4.53	9.2	6,83
Rape distilled	18.9	17.7	4.04	4.12	4.16	3.89
Sunflower distilled	79.0	68.5	4.07	4.22	1.31	1.43
Rape (67%) Sunflower (33%)	2.5	17.6	4.23	4.48	7.24	5.22
Used frying oil	9.3	16.9	4.67	4.94	7.98	5.83
Rape	5.8	9.4	4.60	4.49	7.75	7.00
Rape + PYRO 250 mg/kg	6.9	7.1	4.55	4.50	22.42	20.85
Tallow	n/d	22.0	4.73	5.00	0.70	n/a
EN 14213	n/d	n/d	3.5–5.0	–	6 (min)	–

TBHQ, tert-butyl hydroquinone; PYRO, pyrogallol.

resistant to oxidation, whereas Jatropha methyl esters contain mainly oleic (C18:1) and linoleic (C18:2) acids. Thus, a mixture of the two esters will be more resistant to oxidation (Sarin *et al.*, 2007).

Lubricity

In contrast, unsaturated fatty acids give better lubricity and cold flow characteristics than saturated fatty acids (Knothe, 2005). Desulfurization of conventional diesel leads to a considerable loss in lubricity, which is required for the functioning of the engines' pumps and injectors. A number of studies have shown that the addition of small quantities (5%) of biodiesel can increase the lubricity of conventional diesel (Hu *et al.*, 2005; Knothe, 2005). The polar elements in biodiesel help to form a layer on the metal parts of the engine, reducing engine wear. Lubricity is measured by the ball-on-cylinder lubricity evaluator (BOCLE) test or the high frequency reciprocating rig (HFRR) test and those fuels with good lubricity give values of 4500–5000 g. Figure 7.19 shows the effect on lubricity of adding biodiesel to conventional diesel.

Therefore, the fatty acid content of biodiesel has to be a compromise between the best cetane and oxidative stability and cold flow characteristics. It has been suggested that the ideal fatty acid composition for biodiesel would be 10% myristic (C14:0), 50% palmitoleic (C16:1) and 40% oleic (C18:1) acids. This composition is perhaps the aim of all forms of plant breeding or some form of blend of oil.

Sources of Biodiesel

Biodiesel has been produced using a very wide range of plant oils and animal fats, waste cooking oils and soapstocks. Table 7.17 gives some of the plant oils that have been converted into biodiesel and Table 7.18 the animal fats, waste cooking oil and soapstocks.

All these sources of oil used for biodiesel production have characteristic and different fatty acid profiles and these can influence the properties of the subsequent biodiesel.

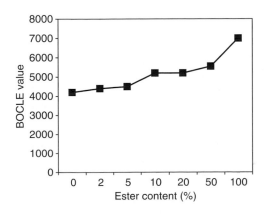

Fig. 7.19. Effect on the lubricity of conventional diesel when biodiesel is added. The lubricity was measured by ball-on-cylinder lubrication evaluator (BOCLE). (Redrawn from Graboski and McCormick, 1998.)

Table 7.17. Plant oils that have been used to produce biodiesel.

Source	Reference(s)
Alternathera triankra	Hosamni *et al.* (2004)
Asclepias syriaca (milkweed)	Hoser and O'Kuru (2006)
Brassica carinata	Bouaid *et al.* (2005); Canoira *et al.* (2006)
Brassica napus (rapeseed, canola)	Nwafor and Rice (1995); Peterson *et al.* (1996); Saka and Kusdrana (2001); Di Serio *et al.* (2005)
Camelina sativa (false flax)	Zubr (1997); Bernardo *et al.* (2003); Frohlich and Rice (2005)
Cotton seed (*Gossypium herbaceum*)	Sahoo *et al.* (2006)
Cynara cardunculus	Enciner *et al.* (2002)
Helianthus annuus (sunflower)	Siler-Marinkovic and Tomasevic (1998); Antolin *et al.* (2002)
Jatropha curcas (jatropha)	Foidl *et al.* (1996)
Jojoba (oil-wax) (*Simmondsia chinensis*)	Canoira *et al.* (2006)
Linseed (*Linum* sp.)	Lang *et al.* (2001)
Lunaria annua (Honesty)	Walker *et al.* (2003)
Madhuca	Ghadge and Rahman (2005); Puhan *et al.* (2005)
Palm oil (*Elaeis quineensis*)	Crabbe *et al.* (2001); Kalam and Masjuki (2002); Ooi *et al.* (2004)
Polanga (*Calophyllum inophyllum*)	Sahoo *et al.* (2006)
Pongamia pinnata (karanja)	Raheman and Phadatare (2004); Karmee and Chadha (2005); Meher *et al.* (2006c)
Rice bran oil	Zullaikah *et al.* (2005)
Rubber seed oil (*Ficus elastica*)	Ramadhas *et al.* (2005)
Soybean (*Glycine max*)	Alcantara *et al.* (2000); Cao *et al.* (2005)
Tobacco seed oil (*Nicotiana tabacum*)	Lang *et al.* (2001); Usta (2005)

Table 7.18. Animal fats, soapstocks and used cooking oils used to produce biodiesel.

Source	Reference(s)
Beef tallow	Ma *et al.* (1998); Nebel and Mittelbach (2006)
Salmon oil	Reyes and Sepulveda (2006)
Used cooking oil	Al-Widyan *et al.* (2002); Tomasevic and Siler-Marinkovic (2003); Ulusoy *et al.* (2004); Cvengros and Cvengrosova (2004); Wang *et al.* (2006)
Used olive oil	Dorado *et al.* (2003)
Vegetable oil soapstock	Hass (2005)
Hazelnut soapstock	Usta *et al.* (2005)
Palm oil fatty acid mixture	Ooi *et al.* (2004)
Tall oil	Altiparmak *et al.* (2007)
Used rapeseed oil	Leung and Guo (2006)
Yellow/brown grease	Canakci (2007)

Glycerol Utilization

Glycerol is a by-product of the production of biodiesel by transesterification, and constitutes 10% of the quantities of the oil used. Rather than discard the glycerol formed, some market needs to be found, as any value obtained from glycerol will go some way to reduce the cost of biodiesel. Initially glycerol was sold to the chemical and soap industries and was used to reduce the overall cost of biodiesel. The effect of glycerol prices on the cost of biodiesel is shown in Fig. 7.20. The rapid growth in biodiesel production has produced a glycerol surplus which has resulted in a drastic decrease in glycerol prices and the closing of glycerol-producing facilities by companies such as Dow Chemicals and Procter and Gamble (Yazdani and Gonzalez, 2007). The price has dropped in the USA from US$0.25/lb in 2004 to US$0.025/lb in 2006. Therefore, alternative uses for glycerol need to be considered.

There are a number of potential uses for the excess glycerol and these are shown in Fig. 7.21. The simplest option is combustion as glycerol has a calorific value (16 MJ/kg). Glycerol can also act as a substrate for anaerobic digestion where it has been shown to stimulate biogas production. It can also be metabolized by some microalgae which can be used as a source of oil or used in anaerobic digestion.

Several microbial species such as *Klebsiella pneumoniae* and *Citrobacter freundii* can ferment glycerol to produce 1,3-propanediol (Mu *et al.*, 2006; Yazdani and Gonzalez, 2007). 1,3-propanediol is used to manufacture polymers (polyesters), cosmetics, foods and lubricants. Another option is to convert the glycerol by etherification (Fig. 7.21) (Karinen and Krause, 2006) which is butoxy-1,2-propanediol which is another fuel. Etherification with isobutene in the presence of an acidic ion exchange resin produces butoxy-1,2-propanediol which has a high octane number (122–128). Thus, this can be an alternative to methyl-tert-butyl ether (MTBE) which is used as an oxygenate. Glycerol has been shown to act as a substrate for citric acid synthesis with *Yarrowia lipolytica* (Papanikolaou *et al.*, 2008). Glycerol has been blended with petrol using a third liquid, ethanol or propanol, to make the two miscible (Fernando *et al.*, 2007). Finally, glycerol can be used as a substrate for the microbial production of plastics poly(3-hydroxybutyrate) PHB and poly(hydroxyalkanoates) PHA (Ashby *et al.*, 2004).

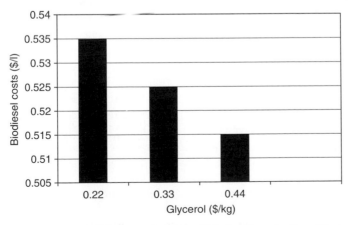

Fig. 7.20. The effect of glycerol prices on the cost of biodiesel production. (Redrawn from Hass, 2005.)

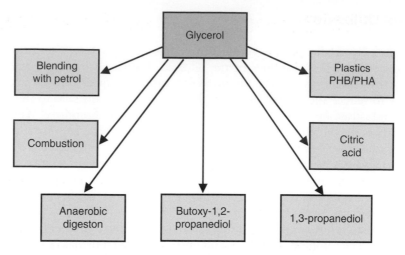

Fig. 7.21. Possible uses for glycerol produced from the transesterification of oil.

Conclusions

Biodiesel produced from plant oils, animal fats and waste cooking oils has been shown to be suitable for use in diesel engines and is currently being added to diesel as a 5% addition. The amount of biodiesel produced worldwide and in the EU is increasing but there are insufficient plant oils available to increase this addition to much more than 10%. If there is any more than a 10% addition, oil crops will begin to compromise food crops (Chapter 8, this volume). Therefore, the other sources of diesel replacements will have to be commercialized such as FT diesel, pyrolysis bio-oil and microalgae.

FT diesel is being produced from the fossil fuels coal and natural gas in two large plants. The plants have to be large to be economic, and even at the large scale, the fuel produced by the FT process is 2–4 times the cost of diesel. To be sustainable, the FT process needs to be run using biomass or waste materials and not fossil fuels. The large scale of the operation would introduce problems of transporting large quantities of biomass to these plants which would burn fuel. What is needed is an improvement in the costing of the FT process which is probable in gas cleaning and in the development of smaller economic units which could treat biomass locally rather than transport it many miles. Pyrolysis is somewhat simpler than the FT process but it would compete for biomass, and bio-oil needs processing before it can be used.

The last option is oil extracted from microalgae, which is at the development stage and so it is difficult to give costing accurately. However, microalgae as a source of biodiesel have a number of advantages over other plant-derived oils. Microalgae can be grown on non-agricultural land, can be grown in sea water, are more productive than land plants and can be part of a CO_2 sequestration system. Thus, they would appear to hold great promise for the future and there is considerable interest worldwide in microalgae.

8   The Benefits and Deficiencies of Biofuels

Introduction

Biofuels are energy sources derived from biological materials and are therefore renewable and sustainable, and can go some distance in replacing fossil fuels and reducing carbon dioxide emissions. Their biological nature separates them from other renewable energy sources such as wind, wave and solar power. Biofuels can be solid, liquid and gaseous and can be used to generate electricity and as transport fuels. No matter how biofuels are used, they have both benefits and shortcomings, and in this chapter these are explored.

The benefits of biofuels whether globally or to a single country are as follows:

1. Reduction in crude oil use. Liquid biofuels can supplement or replace petrol and diesel, and at low levels of blending, little engine modification is required. Biodiesel can be used up to 100% in a conventional diesel engine but higher blends of ethanol (85%) require either modifications or a flexible fuel engine. Biomass and biogas can reduce fossil fuel use for electricity generation.

2. Improvements in engine performance. Ethanol has a very high octane number and has been used to improve the octane levels of petrol. It is also a possible replacement for methyl tertiary butyl ether (MTBE) which is being phased out as an octane enhancer. Biodiesel addition will enhance diesel lubricity and raise the cetane number.

3. Air quality. Biofuels can improve air quality by reducing the emission of carbon monoxide (CO) from engines, sulfur dioxide and particulates (PM) when used pure or in blends.

4. Reduction in the emission of the greenhouse gases (GHGs) carbon dioxide and methane. The replacement of fossil fuels with biofuels can reduce significantly the production of carbon dioxide, and the use of biogas reduces methane emissions.

5. Toxicity. Biofuels are less toxic than conventional fuels, sulfur-free, and are easily biodegradable.

6. Production from waste. Some biofuels can also be made from wastes, for example, used cooking oil can be used to make biodiesel.

7. Agricultural benefits. Biofuel crops of all types will provide the rural economy with an alternative non-food crop and product market.

8. Reduction of fuel imports. By producing fuels in the country, imports will be reduced and the security of energy supply will be increased.

9. Infrastructure. No new infrastructure is required for the first- and second-generation liquid biofuels and some of the solid and gaseous biofuels.

10. Sustainability and renewability. Biofuels are sustainable and renewable, as they are produced from plants and animals.

However, there are shortcomings to the use of biological materials to replace fossil fuels which are as follows:

---

1. Biological material may not be able to produce enough fuel to replace fossil fuels completely, and extensive cultivation of biofuel crops will compete with food crops, perhaps driving up prices.

2. Large amounts of energy are required to produce some biofuels, giving them a low net energy gain.

3. Some of the second- and third-generation biofuels will require the introduction of a completely new infrastructure, for example, hydrogen.

Reduction in Fossil Fuel Use

The present and possible future replacements for fossil fuels have been described in Chapters 4, 5, 6 and 7. The biofuels can be solid, gaseous and liquid as follows:

Solid:

- Biomass.
- Waste materials.

Gaseous:

- Methane.
- Hydrogen.
- Dimethyl ether (DME).

Liquid:

- Methanol.
- Ethanol.
- Biobutanol.
- Synthetic Fischer-Tropsch (FT) petrol.
- Biodiesel.
- Bio-oil.
- Synthetic FT diesel.
- Microalgal biodiesel.

Solid fuel replacements

As described in Chapter 4, biomass in the form of wood, specific energy crops, crop residues and organic wastes can be used to replace coal and natural gas. Biomass can be burnt, co-fired with coal and gasified to generate electricity and heat. Small- to medium-size heating systems have been developed to use pelleted biomass. The pyrolysis and gasification of biomass can be used to produce a liquid fuel, which can be used for transport, electricity generation and to provide heat.

Gaseous fuel replacements

Biogas, hydrogen and DME have been proposed as gaseous replacements for fossil fuels used in transport and electricity generation.

Gaseous fuels have problems of storage and supply not encountered with either solid or liquid fuels. Storage of gas at atmospheric pressure is not practical, and so the gas has to be either compressed at high pressure or liquefied at low temperatures to reduce its volume. Compression to pressures of 200 bar and liquefaction, which for hydrogen requires a temperature of $-253°C$, expends a considerable amount of energy, and subsequent storage has to be in strong pressure vessels or in well-insulated tanks. The lower energy density of the gaseous fuels compared with liquid fuels means that larger fuel tanks are required in vehicles. There is a small number of modified internal combustion engines using gases derived from fossil fuels such as liquid natural gas (LNG), liquid petroleum gas (LPG) and compressed natural gas (CNG). After treatment biogas is the same as natural gas, and therefore could be used as a replacement for liquid and CNG. Biogas has been used as a fuel for boilers, dual fuel engines and the generation of electricity. DME has a boiling point of $-24.9°C$ and so can be liquefied and stored easily. DME has lower energy content than diesel, 28.6 MJ/kg compared to 38–45 MJ/kg, but it has a higher cetane value of 55–60 compared with diesel at 40–55. DME could be used as a replacement for diesel but its properties are similar to propane and butane so that it could also replace these fuels for distributed power generation, heating and cooking. Hydrogen is a high energy clean fuel producing no carbon dioxide on combustion and has been used as a fuel for internal combustion engines and fuel cells. There are problems with the sustainable production of hydrogen, its storage and distribution which may require a completely new infrastructure. The number of alternative-fuelled vehicles in the USA from 1993 is shown in Fig. 5.10. The number of CNG and liquefied natural gas vehicles has remained static, whereas the numbers of liquefied petroleum gas vehicles have declined. Thus, there must be some doubt about the introduction of gaseous biofuels as transport fuels.

Petrol replacements

Ethanol can be used in petrol engines up to a concentration of about 24%, any more and either engine modifications are required or a flexible fuel engine is required. However, with modifications, a petrol engine can run on 100% ethanol, although it contains less energy than petrol. Biobutanol is a possible replacement for ethanol, as it has a higher energy density but there is little information about its use in engines. FT petrol is essentially the same as petrol and should cause no problems in supply and use. Methanol has been used in petrol engines, but it is not used as a fuel at the moment but could be a very useful fuel in the supply of hydrogen to fuel cells.

Diesel replacements

Biodiesel can be used up to 100% in a conventional diesel engine and jet engine without modification, but bio-oil requires considerable processing before it can be used because of its high viscosity and acidic nature. FT diesel is the same as diesel and can be used without modification. The microalgal biodiesel appears to have similar properties to biodiesel, but so far there is no information on its use in diesel engines.

Gas turbine engines are used in aircraft, marine propulsion and electricity generation and can function on a variety of fuels, paraffin, natural gas and propane, but the formulation for aircraft engines is regulated by international specifications DEF

STAN 91-91 and ASTM D1655. Jet A and Jet A-1 are known as aviation kerosene, and Jet B is a wide cut fuel. Biodiesel has been used at concentrations of 2, 20 and 30% in jet fuels to power aviation turbine engines with no apparent ill effects (Wardle, 2003). However, the biodiesel results from diesel engines cannot be transferred directly to aircraft engines as the combustion system responds differently, influenced by density more than chemistry (Ebbinghaus and Weisen, 2001). The problems with the supply of first-generation biofuels may preclude the use of biodiesel in aircraft but it is another option.

Thus, biofuels are capable of supplementing or replacing fossil fuels for transport and electricity generation and could be integrated into the present fuel infrastructure. The only exception is hydrogen which may require a completely new infrastructure.

Fuel Economy with Biofuels

There have been a number of studies on fuel economy and emissions concentrating on biodiesel and bioethanol. Biofuels have different energy content from the fuels that they replace, and therefore this will affect their fuel economy. Fuel economy can be measured in terms of volumetric fuel consumption, brake-specific fuel consumption (BSFC, kg/kWh fuel flow/power) and brake thermal efficiency by measuring the torque and fuel consumption. In other studies, fuel consumption was determined from the carbon dioxide emissions and fuel carbon content but a more accurate value can be obtained when carbon dioxide emissions and fuel consumption are combined. The fuel consumption is normally measured at various power outputs (kW) and loads (Nm).

Biodiesel

The effect of the concentration of biodiesel used in blends on BSFC measured as MJ/kWh is shown in Fig. 8.1. Although the histogram shows a drop in BSFC as the

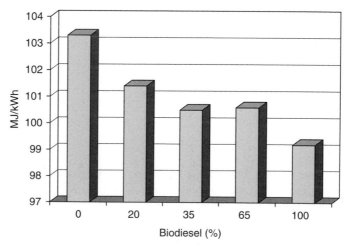

Fig. 8.1. Fuel economy measured as brake-specific fuel consumption (BSFC) using soybean oil biodiesel in a four-stroke DDC series 60 diesel engine. (Redrawn from Graboski and McCormick, 1998.)

biodiesel concentration increases, the final value is only reduced by 4%, and with a standard deviation of 1.5%, it appears that fuel composition has little effect on fuel consumption. The type of oil used to produce biodiesel has also been investigated, and little difference can be seen in the fuel consumption using a variety of biodiesel (Fig. 8.2). This is perhaps not surprising as the calorific value of biodiesel made from various oils varies very little.

The change in engine load will also affect fuel consumption. A number of studies have been carried out and Fig. 8.3 shows a comparison of diesel with a 50% mixture of sunflower oil biodiesel blended with marine diesel tested in a single cylinder,

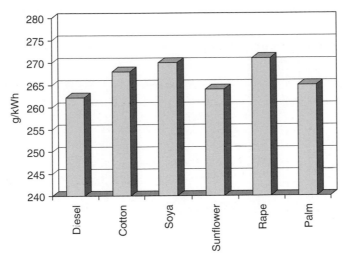

Fig. 8.2. The effect of the oil used to produce biodiesel on the brake-specific fuel consumption (BSFC). The biodiesel was mixed at 20% and used in a single cylinder, four-stroke direct injection diesel engine. (Redrawn from Rakopoulos *et al.*, 2006.)

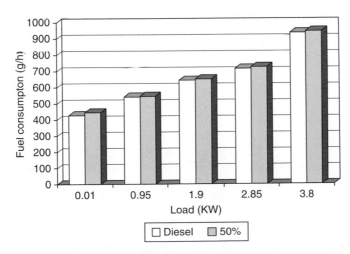

Fig. 8.3. Fuel consumption (g/h) for marine diesel and a blend (50%) with sunflower biodiesel at various loads in a single cylinder diesel engine. (From Kalligeros *et al.*, 2003.)

indirect injection diesel engine. In the case of the marine diesel when compared with a blend of 50% sunflower biodiesel, there was little difference in BSFC over the loads imposed apart from an increase in fuel consumption from around 400 g/h to over 900 g/h. In a similar study on the effect of load on BSFC using a biodiesel made from waste olive oil compared with diesel and oil, there was an increase of 7.5% in fuel consumption compared with diesel (Fig. 8.4).

A number of studies using a variety of biodiesel blends and engines have shown that increases in BSFC can vary from 0 to 13.8% fuel consumption increase over diesel (Table 8.1). Biodiesel showed higher fuel consumption of 13.8% measured as BSFC using soy oil biodiesel in a four cylinder, four-stroke turbocharged John Deere engine (Monyem and Van Gerpen, 2001). The same result was found with biodiesel

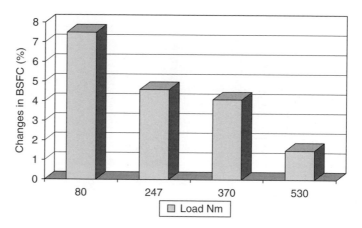

Fig. 8.4. The effect of changing load on the brake-specific fuel consumption (BSFC) using waste olive oil biodiesel in a three cylinder, four-stroke direct injection diesel engine. The results are expressed as changes compared with diesel. (Redrawn from Dorado *et al.*, 2003.)

Table 8.1. Fuel consumption using biodiesel.

Biodiesel source	Mix used (%)	Heating value (MJ/kg)	Engine used	Percentage of reduction in BSFC	Reference
Rapeseed	100	40.15	Petter AC1 single cylinder, direct injection	11% less	Nwafor (2004)
Tobacco seed	10, 17.5, 25	39.8	Four cylinder, four-stroke, direct injection	None	Usta (2005)
Sunflower	10, 20, 50	38.47	Petter marine, single cylinder, indirect injection	None	Kalligeros *et al.* (2003)
Olive oil	10, 20, 50	38.78	Petter marine, single cylinder, indirect injection	None	Kalligeros *et al.* (2003)
Soya	100, 20	37.27	John Deere four cylinder turbo	13.8% less	Monyem and Van Gerpen (2001)
Waste olive oil	100	39.67	Perkins three cylinder, four-stroke direct injection	7% less	Dorado *et al.* (2003)

from waste olive oil run in a three cylinder, four-stroke Perkins diesel engine where the BSFC was 7% greater (Dorado *et al.*, 2003). The reduction in fuel consumption may be a consequence of the lower energy density of biodiesel (33–42 MJ/kg) compared with diesel (46 MJ/kg). The variation in the reduction in BSFC may also be due to the different combustion conditions found in the different engines.

Ethanol

Ethanol has an energy content of 23.5 MJ/l and therefore contains 32.5% less energy than petrol. The most common use of ethanol in the USA is as an E85 (85% ethanol) blend where the energy content is 71.95% of petrol. This should result in 5–12% decrease in fuel economy. A study of the 2007 EPA data estimates that the mean combined fuel economy is 73.4% (Roberts, 2007), although in flexible fuel cars there was little difference in fuel economy between E85 and petrol. The performance of a petrol blend with either 20% ethanol or 20% methanol has been compared with unleaded petrol. The results in terms of maximum brake torque (power) at different engine speeds are given in Fig. 8.5. Both ethanol and methanol improved the brake torque over the unleaded petrol which in both cases was probably due to better anti-knock characteristics and increased volumetric efficiency due to higher oxygen content (Agarwal, 2007). The fuel economy and power generation has not been reported for the other biofuels.

Emissions from Biofuels

One of the advantages of biofuels is the possible reduction in engine emissions which contribute to global warming and atmospheric pollution. Here again most of the studies have been carried out with the first-generation biofuels, ethanol and biodiesel.

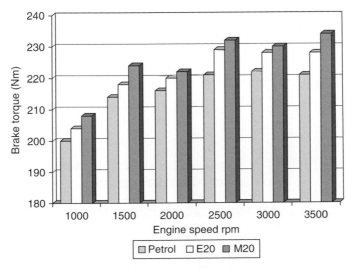

Fig. 8.5. Effect of the addition of 20% ethanol (E20) and 20% methanol (M20) to unleaded petrol on brake torque at various engine speeds. (From Agarwal, 2007.)

Biodiesel

There have been a large number of studies on the exhaust emissions from engines using a variety of biodiesel types and concentrations (Graboski and McCormick, 1998; Willianson and Badr, 1998; EPA, 2002). However, it is difficult to compare results as different engines, conditions, and blends have been used. Figure 8.6 shows the mean of a number of studies on the effect of using 100% rapeseed biodiesel on the important engine emissions: hydrocarbons (HC), CO, nitrous oxides (NO_x), and PM. Rapeseed biodiesel is the main biodiesel produced in the EU and the consensus shows a considerable reduction in the emission of HC and PM and a small increase in NO_x. The increase in NO_x was probably due to an increase in combustion temperature. The mean of the three studies on the effect of sunflower biodiesel on engine emissions is shown in Fig. 8.7. With sunflower biodiesel, the reduction in HC was the

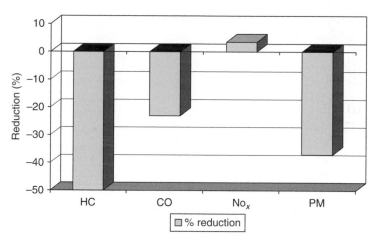

Fig. 8.6. Mean values for the effects of rapeseed biodiesel on the emissions from a number of engines. (From Peterson *et al.*, 1996; Williamson and Badr, 1998; Makareviciene and Janulis, 2003; Labeckas and Slavinskas, 2006.)

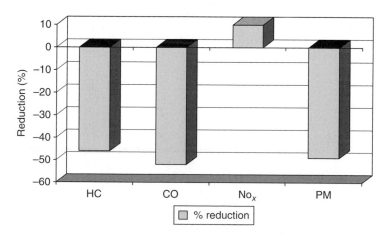

Fig. 8.7. Mean values of the effects of sunflower biodiesel on the emissions from a number of engines. (From Monyem and van Gerpen, 2001; EPA, 2002.)

---

same as rapeseed biodiesel, and CO and PM were further reduced, but NO_x emissions increased. The emissions from an engine fuelled with 100% waste olive oil biodiesel at different loads are shown in Fig. 8.8 (Dorado *et al.*, 2003). As the load increases the reduction in CO, NO_x and sulfur dioxide decreases to zero at the highest load. A different result was observed when a 50% sunflower biodiesel blend was used in a marine diesel engine (Fig. 8.9) (Kalligeros *et al.*, 2003). With 50% sunflower biodiesel, the emissions decrease as for waste olive oil biodiesel but do not reach zero at the highest load. The advantages of using biodiesel to reduce emissions may therefore be eliminated when the engine is used at high loads. However, the reduction in emissions may depend on the test engine used.

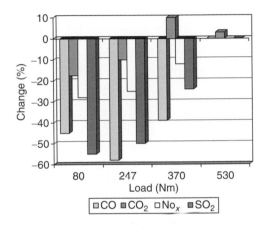

Fig. 8.8. The effect of waste olive oil biodiesel (100%) on the percentage of changes in emissions from a diesel engine compared with diesel at various loads (Nm). (From Dorado *et al.*, 2003.)

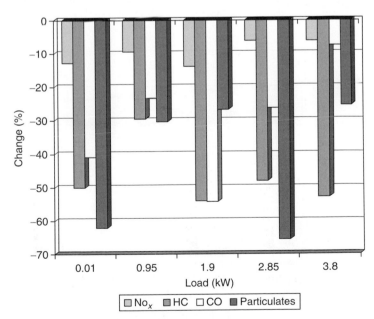

Fig. 8.9. The percentage of change in emissions when 50% sunflower biodiesel blend is used in a marine diesel engine at various loads (kW) compared with diesel. (From Kalligeros *et al.*, 2003.)

The effect of increasing concentrations of biodiesel on engine emissions is shown in Figs 8.10 and 8.11. As the concentration of commercial biodiesel in blends increased, the emission of CO was reduced and NO_x increased (Fig. 8.10). When soybean biodiesel was tested in contrast to commercial biodiesel, CO was not reduced significantly but HC and PM were reduced and NO_x increased.

In general, emissions from diesel engines running on blends or 100% biodiesel showed a reduction in CO, HC and PM, but an increase in nitrous oxide (NO_x) levels. The reason for this change in emissions is thought to be the higher oxygen content of biodiesel, which gives a more complete combustion of the fuel and this reduces CO, HC and PM. The Environmental Protection Agency (EPA) has compiled the results of a number of studies on the effect of biodiesel content on emissions and the results were

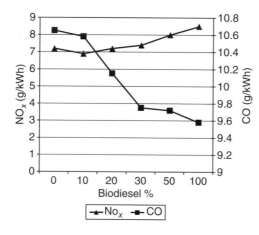

Fig. 8.10. The effect of various concentrations of commercial biodiesel added to diesel in a four-stroke direct injection single cylinder diesel outboard engine. (Redrawn from Murillo *et al.*, 2007.)

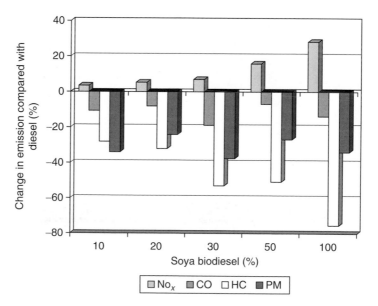

Fig. 8.11. Percentage of change in steady state emissions from a soy biodiesel fuelled Navistar HEUI diesel engine. (From Graboski and McCormick, 1998.)

similar to those observed in Figs 8.10 and 8.11. In a study using a MAN diesel bus engine, the fuel injection characteristics were different for diesel and rapeseed-derived biodiesel (Kegl, 2008). The biodiesel when injected into the engine forms a longer and narrower spray than mineral diesel, caused by a higher injection pressure, increased by low fuel vaporization and atomization due to higher surface tension and viscosity.

The reasons for the increased NO_x production when using biodiesel may be the higher combustion temperature and injection characteristics. The increase in nitrous oxide (NO_x) is probably due to the raised combustion temperature which is known to increase NO_x formation. Advanced injection is caused by the higher bulk modulus of compressibility of biodiesel which allows the pressure wave from the pump to the nozzle to speed up, therefore advancing the timing. It has been observed that retarding the timing can in some way reduce NO_x emissions. In order to reduce the emission of NO_x with biodiesel, the injection timing was altered and the optimum setting was found to be 19° (°CA BTDC, degree of crankshaft angle before top dead centre) compared with 23° for diesel. The effect of altering the injection timing on emissions of CO and NO_x is shown in Fig. 8.12. The lowest NO_x emission was obtained at 21°, but the lowest CO emission was at 24°. In all the studies on emissions, no evidence has been given that the engines were optimized for biodiesel, and therefore modifications such as altering the timing may reduce emissions of NO_x.

Another way of reducing NO_x production is to use exhaust gas recycling (EGR). Diesel engines fuelled with Jatropha oil biodiesel produce more NO_x than diesel (Pradeep and Sharma, 2007). In this case exhaust gas recirculation was tested as a system to reduce NO_x. The exhaust gases consisting of carbon dioxide and nitrogen are recirculated and injected into the engine inlet, reducing the oxygen concentration and combustion temperature which reduces NO_x. The level of recirculation is critical because if the oxygen is reduced too far, incomplete combustion will produce higher levels of hydrocarbon, CO and smoke. In this case with a single cylinder diesel engine, the optimum recirculation was 15% as can be seen in Fig. 8.13.

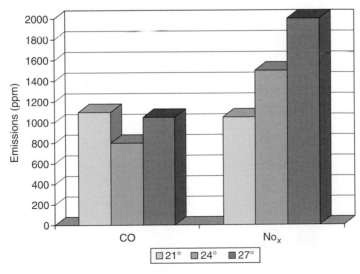

Fig. 8.12. Effect of injection advance on emissions, the normal setting for diesel is 23°. (From Carraretto *et al.*, 2004.)

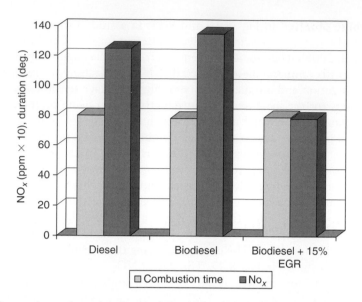

Fig. 8.13. Comparison of combustion duration (degrees) and NO$_x$ emissions for *Jatropha* sp.-derived biodiesel and diesel using exhaust gas recirculation. (From Pradeep and Sharma, 2007.)

Whatever biodiesel is used the scale of reduction in emissions will also be dependent on the engine characteristics such as combustion chamber design, injector nozzle, injection pressure, air–fuel mixture, load and other features. Therefore, the reduction in emission will vary from one diesel engine design to another.

Ethanol

Small quantities of ethanol (3–6%) have been added to petrol to increase the oxygen content to ensure complete combustion and reduce the emission of HC and PM. When high concentrations of ethanol are used such as the E85 fuel in a standard petrol engine, CO and NO$_x$ are reduced compared with petrol but there is an increase in HC (Fig. 8.14). In the flexible fuel engine where the conditions are optimized for E85 fuel, emissions of CO and NO$_x$ were reduced but HC and methane were increased.

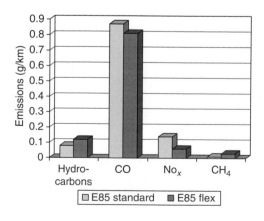

Fig. 8.14. The effect of E85 on the emissions from a standard engine and a flexible fuel engine. CO, carbon monoxide; NO$_x$, nitrous oxides, CH$_4$, methane. (From Wang *et al.*, 1999.)

Dimethyl ether (DME)

A number of studies have been carried out on the emissions from a compression ignition engine (diesel) running on DME and DME blends. DME has been shown to produce low noise, smoke-free combustion and reduced NO_x when used in an internal combustion engine (Huang *et al.*, 2006). DME, because of its high cetane number and low boiling point, has been used at 100% or as an oxygenated addition to diesel. When DME was used in a diesel engine, it reduced NO_x and SO_x emissions and was sootless (Semelsberger *et al.*, 2006). Large motor manufacturers are developing truck and bus transport fuelled by DME. The emission levels from these development vehicles when run on DME show virtually no PM and low levels (0.5–2.0 g/kWh) of NO_x.

Reduction in Carbon Dioxide Emissions when Using Biofuels

Biomass

The carbon dioxide fixed during photosynthesis is released when the biological material is burnt which means that there is no net gain in atmospheric carbon dioxide. This indicates that biological materials are 'carbon neutral' in nature and therefore ideal for the mitigation of global warming. However, one of the main arguments against biofuels of all types is that fossil fuels are used and carbon dioxide released during the production of biofuels. Therefore, biofuels are not 100% carbon-neutral. Clearly fossil fuel will produce the greatest amounts of carbon dioxide as they release carbon dioxide fixed millions of years ago, whereas biomass has fixed its carbon dioxide in the last 10 years. Table 8.2 gives some values for the carbon dioxide generated per megajoule (MJ) of energy during the combustion of fossil fuels. Coal and coke have the highest carbon content and produce the highest levels of carbon dioxide. Natural gas (methane) has the lowest GHG emissions of the fossil fuels and is one of the reasons why electricity generation was switched to gas in the UK in the 1990s. To be suitable for carbon dioxide mitigation, biofuels will have to have GHG emissions much lower than those for fossil fuels.

The bioenergy crops and waste biological materials described in Chapter 4 if used for energy will clearly reduce the amount of carbon dioxide accumulating in the atmosphere. The carbon dioxide emissions from biomass crops were compared with those emitted from the solid fossil fuels coal and coke (Fig. 8.15). The carbon dioxide produced per megajoule when various biomass sources are burnt or used to generate electricity are compared with two solid fossil fuels coal and coke in

Table 8.2. Greenhouse gas emissions from fossil fuels.

Fuel	GHG emissions (g CO_2/MJ)	Reference
Coal	107.1–110.4	Gustavsson *et al.* (1995); Matthews (2001)
Coke	117.0–134.0	Gustavsson *et al.* (1995); Matthews (2001)
Fuel oil	81.3–81.4	Gustavsson *et al.* (1995); Matthews (2001)
Diesel	77.6–81.9	Gustavsson *et al.* (1995); Matthews (2001)
LPG	73.6–80.8	Gustavsson *et al.* (1995); Matthews (2001)
Natural gas	66.2–68.5	Gustavsson *et al.* (1995); Matthews (2001)

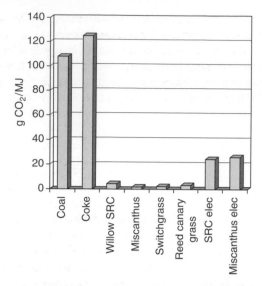

Fig. 8.15. Carbon dioxide emissions from bioenergy crops when burnt or used to generate electricity. SRC, short rotation coppice. (From Gustavsson *et al.*, 1995; Dubisson and Sintzoff, 1998; Matthews, 2001; Bullard and Elsayed *et al.*, 2001; Heller *et al.*, 2001; Keoleian and Volk, 2005.)

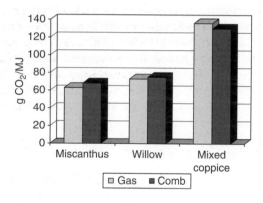

Fig. 8.16. Greenhouse gases saved when biomass is either gasified or combusted in g CO$_2$ equivalents/MJ. (From Lettens *et al.*, 2003.)

Fig. 8.15. The combustion of the biomass produces 20 times less carbon dioxide than coal. When short rotation coppice (SRC) and Miscanthus are used to generate electricity more carbon dioxide is formed per unit of energy than simple combustion.

Biomass can also be gasified and the gas used as fuel for gas turbines to produce electricity. Combustion and gasification as a source of energy have been compared using three biomass sources in terms of the amounts of carbon dioxide saved (Fig. 8.16) (Lettens *et al.*, 2003). The perennial grass *Miscanthus* sp. gives the least carbon dioxide and mixed coppice the greatest, and there appears to be little significant difference between combustion and gasification.

Ethanol

Of all the first-generation biofuels, ethanol requires the most processing. As a consequence large quantities of carbon dioxide are produced and energy used during harvesting and preparation. For this reason bioethanol has been of most concern when the first-generation fuels have been evaluated in terms of energy use and carbon dioxide emission. The substrate used to produce ethanol has a considerable

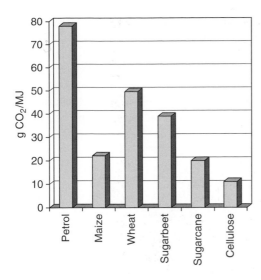

Fig. 8.17. Mean values for carbon dioxide emissions (g CO_2/MJ) from bioethanol production using a variety of substrates. (From Elsayed *et al.*, 2003; Gielen and Unander, 2005; Kim and Dale, 2005; Farell *et al.*, 2006; Hill *et al.*, 2006.)

influence on the energy input and the carbon dioxide released during its production. The carbon dioxide released per megajoule of energy for a number of substrates is shown in Fig. 8.17. Compared with petrol, ethanol from all substrates produces less carbon dioxide with the greatest reduction when produced from cellulose and production from wheat showed the least reduction in carbon dioxide.

Biodiesel

The carbon dioxide produced during the synthesis of diesel and biodiesel combined with carbon dioxide produced when the biodiesel is burnt is given in Fig. 8.18. Diesel produces around 80 g CO_2/MJ compared with 43.7 g CO_2/MJ for rapeseed biodiesel which is a reduction of 45%. In the case of diesel produced by the FT process by the

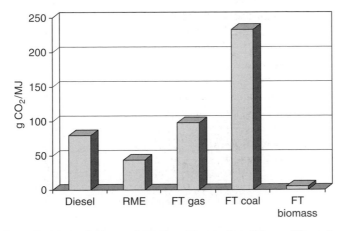

Fig. 8.18. Carbon dioxide emissions when diesel is produced from different substrates. RME, rapeseed methyl ester; FT gas, FT coal and FT biomass: FT diesel produced directly from natural gas and by the gasification of coal and biomass, respectively. (From Mortimer *et al.*, 2003; Gielen and Unander, 2005; IEA, 2005a.)

gasification of coal, natural gas and biomass, the carbon dioxide produced varies considerably. Both natural gas and coal FT diesel produce more carbon dioxide than diesel, 98 g CO_2/MJ and 233 g CO_2/MJ, respectively. The use of biomass in the synthesis of FT diesel yields only 5 g CO_2/MJ which represents a 94% reduction in carbon dioxide compared with mineral diesel.

The carbon dioxide fixed during growth and its distribution in products during the production of biodiesel from rapeseed is shown in Fig. 8.19. The rapeseed plant

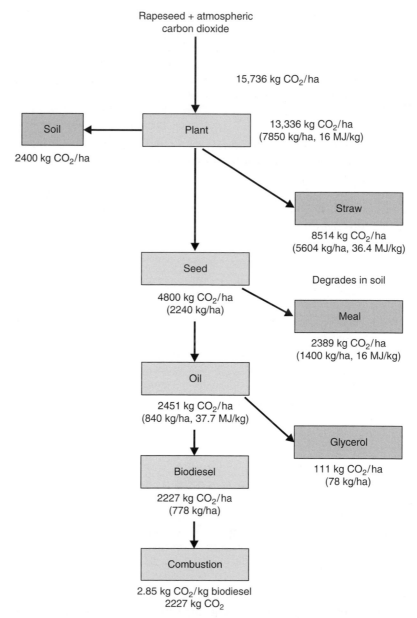

Fig. 8.19. The distribution of carbon dioxide in the production of biodiesel from rapeseed. (From Peterson and Hustrulid, 1998; Mortimer *et al.*, 2003.)

fixes a total of 15,736 kg CO_2/ha, where 13,336 kg was retained in the plant and the remaining 2400 kg stayed in the soil sequestered by the soil microorganisms. The yield of the seed was 2240 kg/ha, 28.5% of the total plant material, containing 4800 kg carbon dioxide/ha. The rest of the plant, the straw, contains 8514 kg CO_2/ha, 63.8% of the carbon dioxide fixed, which is often ploughed into the soil. The oil extracted from the seed contains 840 kg/ha which is an oil yield of 37.5% leaving the meal or cake containing 2389 kg CO_2/ha to be used as feed or fuel. On transesterification the oil is converted into 840 kg biodiesel and 78 kg glycerol. The biodiesel represents some 9.2% of the whole plant and yields 2227 kg CO_2/ha when used as a fuel. Figure 8.19 also includes the energy content of the biodiesel and the co-products where it can be seen that the meal and straw contain more energy than the biodiesel. For these reasons, it may be more efficient to convert the whole plant or biomass into a biofuel rather than just the oil extracted from the seed. This should be the nature of the second-generation biofuels.

The unit operations used in the production of biodiesel have been evaluated for their carbon dioxide and GHG production and energy input (Fig. 8.20a,b,c). It is clear that the major carbon dioxide-producing stages were esterification, use of nitrogen fertilizer and to a lesser extent solvent extraction of the oil. A second process has been included where solvent extraction has been replaced by cold pressing and low nitrogen growing conditions. This results in more than 50% reduction in energy input, carbon dioxide and GHG emissions. Thus, the processes of biofuel production can be made more environmentally suitable by modifications to the key stages of the process.

Gaseous and second-generation biofuels

The gaseous fuels DME and hydrogen can be produced using a number of routes including those from biological materials. The carbon dioxide produced per unit of energy (g CO_2/MJ) is shown in Fig. 8.21 for hydrogen and DME. The production of DME from syngas, hydrogen from electrolysis, natural gas and coal is compared with petrol, CNG, LPG, coal and gas in terms of carbon dioxide produced per unit of energy. The amount of carbon dioxide produced by DME is similar to petrol and LPG. Hydrogen production by all three routes produces more carbon dioxide than petrol especially when coal is used. This indicates one of the problems of producing hydrogen from fossil fuels.

In a study by the Joint Research Centre EU, the cost of reducing carbon dioxide emissions for a number of biofuels was calculated and some of the data is shown in Fig. 8.22. The GHGs avoided in a life-cycle analysis, well-to-wheel (WTW), are related to the cost (€/t) of carbon dioxide equivalents avoided. A life-cycle analysis systematically identifies and evaluates opportunities for minimizing the overall environmental consequences of using resources and releases into the environment. In terms of GHGs avoided, biodiesel is slightly less expensive than ethanol whether produced from either sugarbeet or wheat. The fuels DME, ethanol and FT diesel produced from wood biomass avoid the most GHGs because of using a sustainable source and are inexpensive at around €100–500/t CO_2 avoided. Unfortunately the production of the biofuels from wood has yet to be commercialized. Electrolysis of water to produce hydrogen using sustainable electricity from nuclear and wind power avoids large quantities of GHGs.

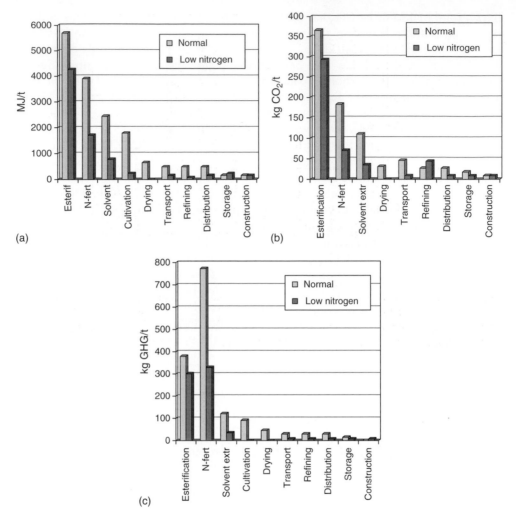

Fig. 8.20. (a) The energy input (MJ/t) for the processes used to produce biodiesel from rapeseed under normal and low nitrogen cultivation. (b) The carbon dioxide (kg/t) produced from the various stages of biodiesel production from rapeseed under normal and low nitrogen conditions. (c) Greenhouse gas (kg/t) for biodiesel production from rapeseed using a conventional and low nitrogen process. (Redrawn from Mortimer *et al.*, 2003.)

CNG, hydrogen from natural gas and ethanol from wheat only avoid moderate amounts of GHGs. Hydrogen for use in fuel cells produced by on-board reforming of petrol is the most expensive option.

Biodegradability of Biofuels

One of the advantages of biodiesel and some of the other biofuels is that they are non-toxic and degrade more rapidly than fossil fuels. This is an important feature in the case of accidents and spillages. Marine environments, freshwater, soil and various

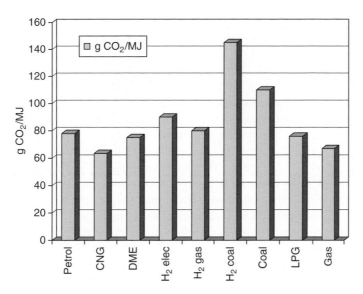

Fig. 8.21. Carbon dioxide produced per unit of energy for gaseous fuels compared with some fossil fuels. CNG, compressed natural gas; LPG, liquid petroleum gas; DME, dimethyl ether; hydrogen produced from electrolysis, natural gas and coal. (From Gustavsson *et al.*, 1995; Matthews, 2001; Gielen and Unander, 2005; IEA, 2005a.)

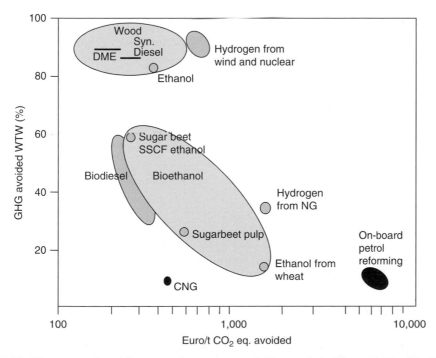

Fig. 8.22. The costs of avoiding greenhouse gas emissions using different fuels. (Modified from JRC, 2007.)

sediments have been contaminated with oil components throughout the world as a result of accidents, leaks, spills and disposal. Oil components can cause considerable environmental disruption and the most spectacular are those accidents involving oil tankers.

One definition of biodegradability is where a fuel is 90% or more degraded within 21 days under fixed conditions (Sendzikene *et al.*, 2007). The determination of degradation can be followed by the reduction in the concentration of the fuel as estimated by analysis such as gas liquid chromatography or the release of breakdown products such as carbon dioxide (Zhang *et al.*, 1998). The two methods give somewhat different results. Using carbon dioxide emissions rapeseed methyl esters degrade by 77–89% in 28 days compared with 18% for mineral diesel. Using gas liquid chromatography (GLC) the values were 88% for rapeseed methyl esters and 26% for mineral diesel (Zhang *et al.*, 1998). This is not surprising as there should be a delay for the fuels to be fully mineralized to carbon dioxide compared with the metabolism of the methyl esters.

More recently the concentrations of oils and fatty acid methyl esters have been estimated using infrared spectroscopy and Fourier-transformed infrared spectroscopy (FTIR). In Europe, the biodegradability of fuel has been measured using infrared with the absorption of the C-M stretch of CH_2—CH_3 at 2930 cm^{-1} (Sendzikene *et al.*, 2007). In other determinations the band at 1573 cm^{-1} (COO^-) was used (Al-Alawi *et al.*, 2006).

The biodegradation of a number of plant oil- and animal fat-derived biodiesel has been determined using a microbial mixture obtained from a wastewater system. The results are shown in Fig. 8.23. Diesel was about 60% degraded by day 21 whereas linseed-, tallow-, lard- and rapeseed-derived biodiesel were over 90% degraded in the same time. The rate of degradation was greatest with tallow- and lard-derived biodiesel.

Mixing biodiesel with mineral diesel increases the rate of degradation of mineral diesel and as the concentration of biodiesel increases so does the rate of degradation (Table 8.3). The reason for the increase in degradation is not known but may be due

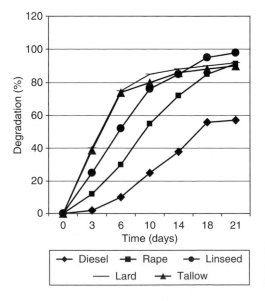

Fig. 8.23. Biodegradation of various biodiesel preparations over 21 days (Redrawn from Sendzikiene *et al.*, 2007.)

Table 8.3. The effect of ester content on degradation rates of mineral diesel. (Adapted from Sendzikiene *et al.*, 2007.)

Mixture	Percentage of degradation
100% rapeseed methyl ester	91
75	91
50	90
25	70
0	58

either to the provision of a more accessible substrate biodiesel or the solubilization of mineral diesel. Biodiesel has been shown to solubilize mineral diesel and has been used to remove crude oil from contaminated sand (Pereira and Mudge, 2004).

Supply of Biofuels

One of the main criticisms of the first-generation biofuels is that they appear incapable of supplying the large quantities of transport fuels required without compromising food crops. The first-generation biofuels were not intended to completely replace fossil fuels but rather to demonstrate that alternative fuels could be used in the internal combustion engine and for electricity generation. Any calculation of the land required to replace 100% fossil fuel would indicate this. It was the development in the second and third generation that was intended to supply the bulk of the fuel. In addition, the introduction of some new fuels may require the installation of a new infrastructure which will take some time to install. The sustainable systems for electricity generation such as wave and wind power can be integrated into the current electricity supply systems, but the introduction of new transport fuels will incur high costs if these fuels are not compatible with current supply infrastructure. To be compatible with present infrastructure, the non-fossil fuels should be liquid as this fits the engine technology and supply systems. Gaseous fuels such as hydrogen and DME will require either the introduction of a new infrastructure or significant modification of the present systems. Once the first-generation biofuels were shown to be suitable for present engine technology their introduction was driven by the legislation produced under the Kyoto Protocol for carbon dioxide reduction.

The primary sources of transport fuels are all agricultural crops such as wheat and maize, apart from wood, woody and organic wastes, and many of these are food crops. Therefore, a conflict may occur between food and fuel crops.

The yield of fuel obtained per hectare varies depending on the crop, growth conditions, and climate. The yields of oil for biodiesel range from 5000 t/ha for oil palm to 1000 t/ha for rapeseed and 375 t/ha for soybean, which is mainly grown for its protein content. A number of studies have been carried out on the effect of biofuel crops on agriculture (Azar, 2005; Johansson and Azar, 2007). The detractors of biofuels have perhaps been too simplistic in their approach to biofuels, and as a consequence biofuels have been blamed for food shortages and increases in food prices (Johansson and Azar, 2007). The adoption of large-scale production of

first-generation biofuels has been recognized as having some unfortunate consequences in addition to their obvious advantages (OECD, 2007). The consequences
are mainly the conflict between the growth of food crops and those for biofuels.
This has been suggested as the reason for the shortage of certain foods and the rise
in the price of others. However, food prices on a large scale are subject to a large
variety of factors so that the reasons for price rises are complex and cannot be
solely due to biofuels.

Producing biofuels on a large scale will require large land areas in countries where
the land is required for food production. In two scenarios produced by the International Food Policy Research Institute (IFPRI), they predict that the price of maize
and oilseed will increase by 26 and 18%, respectively, if the production of biofuels
continues as currently planned. If biofuel production doubles, the price of maize
would increase by 72% and oilseed by 44%. This increase in food prices would affect
the poor populations who spend a higher proportion of their income on food. The use
of non-food-producing land is the obvious solution to this problem but the yields on
poor land will naturally be reduced. If the price of biofuel crops exceeds food crops,
farmers will plant energy crops on normal agricultural ground. The outcome, at least
for the US markets, will be that farm gate prices for all crops will increase substantially
with a doubling of wheat prices. Increases in the farm gate prices are predicted not to
affect food consumption as the price is low compared with commodity prices. However, if biofuels are produced from lignocellulose the increase in the value of lignocellulose crops may stimulate the use of marginal land for non-food crops. Thus, the
development of second and third generation biofuels needs to be pursued with some
urgency to avoid conflict with food crops.

In order to reduce carbon dioxide emissions, carbon tax and trading schemes
have been introduced, for example, €90/t carbon in the EU. However, once biofuel
crops reach a certain price, driven by the carbon tax, commercial growers will use the
most productive land for biofuel crops, replacing food crops. The conclusion was that
at a carbon price of US$70/t carbon energy crops will dominate other agricultural
options (Johansson and Azar, 2007).

To supply anywhere near the total requirement for liquid fuels either globally or
in the UK will require the introduction of a mixture of fuels and propulsion systems
coming from multiple sources rather than a single source. Perhaps one answer would
be to stop using liquid fuels, abandon the internal combustion engine for alternative
power sources such as fuel cells, and electric motors. However, liquid fuels are supported by a vast infrastructure and industries which supply all the various components and employ a large number of people. Any change from our present position
on fuels will need time, legislation and money and should go through a number of
intermediate stages. A fossil fuel such as diesel may still be required for some time, as
it is the motive force for the largest transport systems such as ships and trains, but we
need to act now to mitigate the problem of global warming and fuel supply.

Transport fuel in the USA

The scale of the problem of providing a significant replacement for fossil fuels using
first-generation biofuels is illustrated in Table 8.4. The agricultural land required to
supply 15% of the US transport fuels for a number of crops has been calculated. For

Table 8.4. Agricultural land needed to supply 15% of the transport fuels in the USA. (Adapted from Gressel, 2007.)

Crop	Oil yield (l/ha)	Area needed (million ha)	Percentage of US crop area
Maize (ethanol)	172	462	178
Soybean (biodiesel)	446	178	67
Rapeseed (biodiesel)	1,190	67	42
Jatropha sp. (biodiesel)	1,892	42	13
Oil palm (biodiesel)	5,950	13	7.2
Algae/cyanobacteria[a] (biodiesel)	59,000	1.3	0.72

[a]Oil content 30%.

the temperate crops such as soybean and rapeseed, this would require 67 and 42% of the agricultural land, respectively, and even with the highest yield crop, oil palm, 7.2% of the land would be needed. Unfortunately crops such as oil palm are unsuitable for growth in temperate climates. The table does make a good case for biodiesel from alternative sources such as microalgae.

European biofuel supplies

The International Energy Agency predicts that the world's energy consumption will increase by 1.7% per annum (IEA, 2005a). If this is applied to the UK's transport fuel, the UK would require 63,000,000 t of liquid fuels by 2010 and 67,402,000 t by 2020. However, a recent report forecast that the total use of fuel in the EU 25 (JRC, 2007) will increase up to 2010 and then slowly decline (Fig. 8.24).

Diesel for cars will steadily increase due to their better fuel consumption and emissions while petrol use will decline. The only group which appears to increase over the time period is the diesel demand by heavy goods vehicle (HGV) transport which is a feature of economic growth within the EU. The increase in fuel use from 2005 to 2020

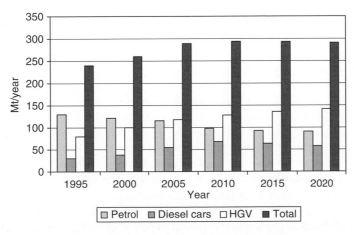

Fig. 8.24. Present and future use of road fuels in the EU 25. (Redrawn from JRC, 2007.)

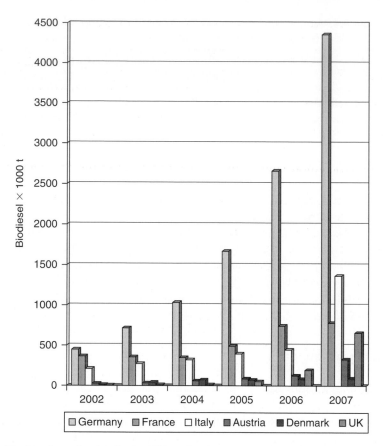

Fig. 8.25. Biodiesel production in the EU 25 for the years 2002–2007 in 1000t. (From European Biodiesel Board, 2007.)

represents an increase of 0.69%, much less than predicted, from global energy use but much of the global growth is probably from developing nations.

The major biofuel produced in Europe is biodiesel and in 2005 Europe consumed 178,178,000t of fossil fuel diesel. The European production of biodiesel is given in Fig. 8.25 where Germany is by far the largest producer, although the UK's production has recently increased significantly. The total biodiesel produced in Europe was 6,069,000t which was considerable, but still only represents 2.45% of the total diesel used.

UK biofuels supply

The UK needs to provide a secure supply of fuel at a time when the UK is increasingly dependent on imports for energy production. At present the UK imports 5% of its gas and 34% of its oil. However, this is changing as the UK continental shelf oil and gas production will reduce over the next 25 years.

The current figure for the use of liquid fossil fuels in the UK (2005 figures) is 58,818,000t (Table 6.1) principally made up of petrol 19,918,000t, diesel 23,989,000t and kerosene for aviation 10,765,000t.

In a recent report the Biofuels Research Advisory Council (*Biofuels in the European Union, a vision for 2030 and beyond*, 2006), the average annual growth for primary energy was predicted to be 0.6% for the UK, which matches the EU 25 prediction. This would include an increase in energy imports from 47.1% in 2000 to 67.5% in 2030. The largest increase will be in fuel for heavy transport.

It is not an impossible task for the UK, to provide 5% of the total diesel used as biodiesel as this would require 6.6% (1.1 Mha) of the total agricultural land set down to rapeseed (Fig. 8.26). In addition, there are 50–90 million l of used cooking oil which could provide 0.24–0.44% of the total diesel. Other estimates indicate that 1.15 Mha of agricultural land would be required to meet the 5% obligation (Rowe *et al.*, 2009).

In the UK, it is rapeseed that dominates the oilseed production, and Fig. 8.26 shows the land required to produce 5, 20 and 100% of the UK's diesel from rapeseed. It is clear that beyond a 5% substitution, biodiesel from rapeseed will require a significant amount of agricultural land which will impact on food crops. Therefore, other sources of biodiesel will be needed and Fig. 8.26 also shows the land required for two alternative sources of diesel, FT diesel and microalgae.

Ethanol can be produced in the UK from sugar extracted from sugarbeet, starch extracted from wheat and cellulose from lignocellulose material. The yields of ethanol per hectare from sugar, starch and lignocellulose crops vary depending on the crop as can be seen in Fig. 8.27. Crop yields also depend on the crop cultivation and climate conditions. The average values are between 2000–3000 l/ha for sugarbeet but the temperate crops cannot compete with sugarcane at 6000 l/ha and sorghum at 4000 l/ha. However, as these crops cannot be grown in the UK, wheat and sugarbeet have to be used. The amount of land for ethanol production at 5, 20 and 100% substitution is shown in Fig. 8.28. Ethanol production from sugarbeet would require 50–60% of the agricultural land for 100% replacement, wheat 54–298% agricultural land and lignocellulose 42–81% agricultural land.

Fuels from biomass

Both diesel and petrol replacements can be produced from biomass and waste organic materials by gasification followed by FT synthesis. This process yields a mixture of

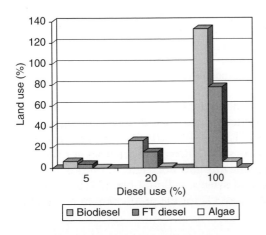

Fig. 8.26. The percentage of agricultural land required to produce 5, 20 and 100% of UK diesel using biodiesel, FT diesel and microalgae. The UK's agricultural land is 18,016,981 ha (18 Mha) and the diesel required is 23,989,000 t. Yield of biodiesel is 1 t/ha, FT diesel 1.35 t/ha and microalgal biodiesel 22.93 t/ha.

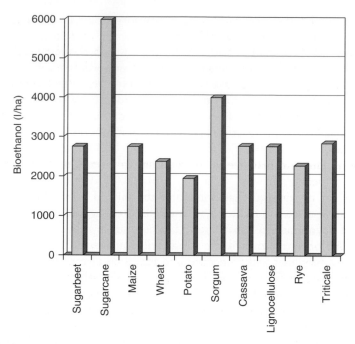

Fig. 8.27. Yields of ethanol per hectare for a number of starch and sugar crops. (From Wheals *et al.*, 1999; Rosenberger *et al.*, 2002; EUBIA, 2005; Dai *et al.*, 2006.)

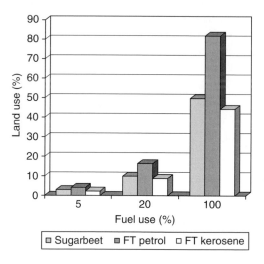

Fig. 8.28. The percentage of agricultural land required to produce 5, 20 and 100% ethanol replacement for petrol from sugar and FT petrol. The petrol use in the UK is 26,916,216,000l (19,918,000t) and the agricultural land is 18,016,981 ha. Ethanol yields of sugarbeet are 2500–3000l/ha; wheat 500–2750l/ha; lignocellulose 1840–3580l/ha. FT petrol and FT kerosene yields are 1.35t/ha.

hydrocarbon which includes diesel and petrol (naphtha) fractions. It has been estimated that FT diesel production is about 45% efficient and therefore the 7,854,422t of woody biomass available in the UK would yield 3,534,489t FT diesel if all the products from the FT process were diesel. Other authors give different efficiencies for the production of liquid fuels from biomass. Van der Drift and Boerrigter (2006) quote that the biomass to syngas efficiency was 80% and syngas to liquid fuels 71%, giving an overall efficiency of 57%, and suggest a value of 63% for the production of liquid fuels if all the products are regarded as diesel.

However, the FT process produces a mixture of HC and the yield of diesel is generally around 30% of the total HC produced. This would represent 1,060,346 t of diesel from 7,854,422 t of biomass. The demand for diesel is 23,989,000 t and therefore FT diesel could provide 5.6% of this demand. Clearly any improvement in efficiency would improve the yield, but at present to increase the amount of FT diesel more biomass crops will need to be planted. In some cases with the FT processes, if the naphtha fraction is recycled it gives an overall yield of diesel of above 50%. The FT production of petrol has the same problems as diesel and petrol constitutes 30–40% of the total product.

The land required to produce FT diesel is shown in Fig. 8.26. The supply of FT diesel assumes a yield of biomass of 10 t/ha, 45% conversion, and 30% of the products are diesel giving a yield of 1.35 t/ha of FT diesel. A 5% addition would require 888,000 ha, 4.9% of the agricultural land, and if 100% replacement was attempted it would use 98% of the present agricultural land. Even if 100% of the FT products were diesel, this would still represent 23–29% agricultural land, a figure which is too high as it would affect food production. The land required providing similar quantities of FT petrol and FT kerosene is given in Fig. 8.28. A 5% addition of FT petrol would need 4.1% of the agricultural land and 100% would require 82% of the land. The amount of kerosene required is less than diesel and petrol and therefore less land is needed with 5% requiring 2.2% of the agricultural land and 100% requiring 44.3% of the land. If all the biomass available in the UK was converted to FT petrol, it would yield 3.5 million t if solely used for petrol production, which is 13.0% of the total use. Thus, FT synthesis of petrol and diesel uses non-food crops but it still requires large land areas to give 100% replacement, which may compromise food crops. Another feature is that the woody biomass available in the UK is widely distributed so that not all of it will be available and it may need transporting some distance.

In addition, another problem with the conversion of biomass to fuels is that to be efficient FT production plants have to be large and it is envisaged that the UK would only need three to four of these units. These large units would involve the transportation of large quantities of biomass and subsequent use of fuel. The development of small, efficient, regional FT plants would be a considerable advance in the provision of FT diesel and petrol.

An alternative to the FT synthesis has been developed where the synthesis gas is converted into ethanol by a bacterial culture (section 'Commercial Lignocellulose Processes', Chapter 6). A company, Bioengineering Resources Inc. (BRI), has developed this system and one unit uses 100,000 t of waste, generates 5–6 MW of power and yields 6–8 million gallons of ethanol (22.68–30.24 million l).

If the UK woody biomass of 7,854,422 t was available, it would supply 78 BRI units yielding 1769–2358 million l representing between 6.6 and 8.7% of the ethanol required for 100% replacement of petrol.

Microalgal biodiesel

Another possible source of biodiesel are the microalgae, which will not grow on agricultural land. Microalgae are more photosynthetically efficient than land plants and the consensus for microalgal production and carbon dioxide fixation are 1.7 g/l/day and 25.7 g CO_2/m²/day, respectively. For example a 10,000 l microalgal bioreactor with a depth of 0.15 m would cover an area of 6670 m² or 0.67 ha. The bioreactor would yield 51,000 kg biomass within a 300-day year. If the oil yield from the microalgae was 30%, this would yield 15,300 kg oil per year which is equivalent to 22,930 kg oil per hectare or 22.93 t/ha which is considerably better than rapeseed at 1 t/ha. In the data put forward by Chisti (2007), the yield of oil was 30,000 kg from 100,000 kg of biomass from an area of 7828 m². This is a yield of 38.32 t/ha which is somewhat higher than the figures used above. Therefore, to replace 100% diesel, 23,989,000 t, using the lower yield figure would require 1,046,184 ha, which represents 5.8% of the agricultural land (Fig. 8.26).

There has been some discussion about the provision of biofuels from microalgae. Chisti (2007) has proposed that biodiesel from microalgae is considerably better than the use of terrestrial crops. In contrast Reijnders (2008) has taken three studies on the life cycle of microalgal fuels and has concluded that the net energy yield of oil palm and sugarcane was better than microalgae (Table 8.5). The table contains raw energy data and net energy yields where the net energy gains for sugarcane and oil palm are better than microalgae, but in terms of raw energy they are very similar.

However, the study does not take into consideration the environmental effects of increasing both sugarcane and oil palm at the expense of rainforest, whereas microalgae can be grown on non-agricultural land, in marine environments and in temperate climates. In a study of oil extraction, sunflower seeds were compared with oil extraction from microalgae (Bastianoni *et al.*, 2007). The results show that sunflower oil extraction was more energy-efficient, but if extraction from microalgae can be improved it should be considered as a fuel source.

Possible biofuel contributions to all transport fuels

From the figures above, it is clear that no single biofuel will be able to fully replace either diesel or petrol but a combination of fuels may be suitable (Table 8.6). The UK has 18,166,000 ha of agricultural land available for biofuel crops. If the first-generation fuels are considered using a 5% addition to petrol and diesel, the total land required would be 9.58% of the agricultural land (Table 8.6). Any more land would begin to conflict with food crops. Therefore, second- and third-generation biofuels have to be considered for biofuel contributions above 10% of the total. The two best candidates are FT biodiesel and FT petrol and microalgal biodiesel, as the FT system uses the

Table 8.5. Energy production by plants and microalgae. (From Reijnders, 2008; Dismukes *et al.*, 2008.)

	Raw energy (GJ/ha/year)	Net energy yield (GJ/ha/year)
Maize	120	–
Sugarcane	1230–1460	161–175
Rapeseed	73	–
Oil palm	–	142–180
Spirulina sp.	550, 1230–1435	127
Tetraselmis suecica	700–1550	–

Table 8.6. The percentage of agricultural land required to produce 5, 20 and 100% fossil fuels replacement using first-, second-, and third-generation biofuels.

Fuel	5%	20%	100%
Biodiesel			
Quantity required (t)	1,199,000	4,797,000	23,989,000
Biodiesel from rapeseed yield 1 t/ha, percentage of land required	6.6	26.6	133
Ethanol			
Quantity of fuel required (t) (percentage of replacement)	995,000 (5)	3,983,000 (20)	19,918,000 (100)
Ethanol from sugarbeet 2.5 t/ha, percentage of land required	2.98	9.9	49.8
FT diesel			
FT diesel yield 1.35 t/ha, percentage of land required	4.9	19.7	98
FT petrol			
FT petrol yield 1.35 t/ha, percentage of land required	4.1	16.4	82
FT kerosene			
FT kerosene yield 1.35 t/ha, percentage of land required	2.2	8.8	44.3
Microalgal biodiesel			
Microalgal biodiesel yield 22.93 t/ha, percentage of land required	0.29	1.16	5.8

whole plant and microalgae non-agricultural land. To provide 20% of both petrol and diesel requirements using FT synthesis would need 36.1% of the agricultural land. This is probably not acceptable even if 9.6% of the land is not used for first-generation biofuels. This suggests that if a large proportion of fossil fuels are to be replaced, fuel generated from non-agricultural land is required. The minimum yields have been used for the FT diesel and petrol but these may be increased through research, and at a yield of 5.7 t/ha, 9.2% of the agricultural could provide 40% of the total fuels. If microalgal biodiesel could provide 40% of the total fuel as biodiesel which would require 2.32% of the land, the total land used would be 11.52% if the microalgae were using agricultural land. This would leave a shortfall of 20% diesel and 60% petrol, but given the assumption that 20% of the fuel use can be saved by increases in efficiency and the continued use of fossil fuels, this shortfall may also be filled. Biodiesel is also being tested as a jet fuel which would reduce the amount of fuel needed from FT synthesis.

Production of Electricity from Biomass

In the UK it has been estimated that there are 7.8 million t of wood biomass available which could be used to produce electricity (Table 4.8, Woodfuel, 2007). The energy content of wood biomass is around 15 GJ/t and therefore 7.8 million t represents an energy content of 0.118 EJ. The UK electricity demand in 2005 was 387.3 TWh (10^{12}) which is 1394 PJ (10^{15} J). Electricity generation is around 30% efficient and therefore

0.118 EJ of energy would yield 35.4 PJ electricity which represents 2.54% of the total requirement. To supply all the electricity would require 92 Gt of biomass which at 12 t/ha represents 7.6 Mha, which is 42.6% of the agricultural land. Another study by Powlson *et al.* (2005) estimated that wood biomass, waste straw and the conversion of some grassland to biomass could yield 12.2% of the UK's electricity. The yields from forestry waste, wheat straw, sugarbeet, set-aside and specifically grown biomass are given in Table 4.9. The woody biomass can only be used once and perhaps it would be better to use the biomass for the FT synthesis of liquid fuels and generate electricity from other renewable sources.

Improvement in Biofuel Quantity and Quality

Any improvement in fuel crop yield in terms of quantity and quality will reduce the amount of land required and processing. It may be possible to improve the quantity of biofuels and change the quality using either conventional plant breeding or genetic manipulation. There is an embargo on genetically manipulated plants within the EU at present, but globally there are a number of countries that have grown large quantities of genetically manipulated crops without any obvious problems. It would appear that the non-food crop would be most suitable for genetic manipulation if regulations were relaxed. Some suggest that genetic modifications are essential if biofuels are to be fully exploited. Some of the areas that genetic manipulation may be helpful in the development of biofuels are as follows:

- Improvement in photosynthesis.
- Improvement in salinity and drought resistance.
- Herbicide and pesticide resistance in fuel crops.
- Improvements in the quantity and quality of plant oils.
- Reduction in lignin in lignocellulose to make cellulose more available for digestion.
- Improvement in hydrogen production by microalgae.
- Improvement in conversion of lignocellulose, cellulose, and other sugars to ethanol.

Improvement in photosynthesis

The surface of the Earth is 510,072,000 km^2 which on average receives 170 Wm^{-2} which is equal to 7500 times the world's energy use of 450 EJ with a total of 3,375,000 EJ. In temperate zones, the amount of energy reaching the Earth's surface is about 1.3 kW m^{-1} but only 5% of this energy is converted into carbohydrates by photosynthesis (Fig. 8.29). Any improvement in the efficiency of photosynthesis would have a considerable effect on crop and biofuel production. The rate-limiting steps in the fixation of carbon dioxide by photosynthesis are ribulose-1,5-bisphosphate carboxylase (rubisco), regeneration of ribulose bisphosphate and the metabolism of triose phosphates. In addition, photosynthetic organisms stop growing and fixing carbon dioxide at light intensities lower than typical levels at midday in equatorial regions. The typical light intensity is about 2000 μmole m^{-2} s^{-1}, whereas light saturation of photosynthesis occurs at 200 μmole m^{-2} s^{-1}. Above the saturating light intensity inhibition of photosynthesis, known as photoinhibition, occurs (Fig. 8.30). Improvements in the three stages of carbon fixation, the increase in the light saturation values and

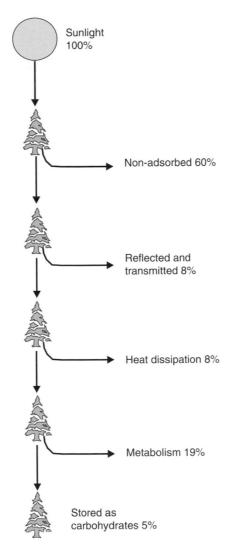

Fig. 8.29. The amount of energy converted into carbohydrates during photosynthesis.

reduction in photoinhibition would all be of value in increasing crop yields. During photosynthesis, leaves must also dissipate heat and improvements in heat tolerance would also be useful.

Salinity and drought tolerance

The major crops for the first-generation biofuels ethanol and biodiesel are rapeseed, soybean, sunflower, oil palm, sugarcane, sugarbeet and maize. The ability to grow these crops in low moisture and saline conditions would relieve the pressure that biofuels have on prime agricultural land.

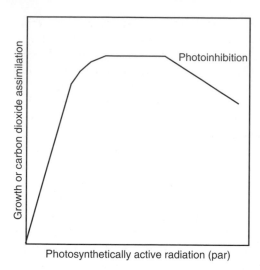

Fig. 8.30. The effect of light intensity on growth and carbon dioxide assimilation by plants.

Y-axis: Growth or carbon dioxide assimilation

X-axis: Photosynthetically active radiation (par)

Label: Photoinhibition

Herbicide, pesticide and pest resistance

In the USA, the main crop for ethanol production is maize and since 1997 genetic manipulation has increased the yield of maize per hectare from 7.5 to 9.0 Mt/ha which is above the increase expected by traditional plant breeding (McLaren, 2005). This has been achieved by engineering maize to be resistant to the European corn borer by inserting a gene from *Bacillus thuringiensis* (Bt) that codes for a protein with insecticide activity. Another example is the pollen beetle which is a pest affecting the pollen of rapeseed, the main European source of biodiesel. There are no natural plants resistant to this pest capable of breeding with rapeseed and so genetic manipulation was the only solution. Transgenic rapeseed plants have been produced carrying a pea lectin gene and the lectin was found to be toxic to the pollen beetle larvae. The pea lectin was placed under the control of a pollen-specific promoter so that the lectin is not present in the rest of the plant (Ahman *et al.*, 2006).

Improvement in quantity and quality of plant and microalgal oils

An increase in the oil content of the main oil-producing plants has obvious advantages, especially with soybean. Soybean is grown as a source of high protein animal feed but supplies large amounts of oil in the USA, although it only contains low levels of oil (18–22%). The limiting step in fatty acid synthesis would appear to be the acetyl-CoA carboxylase production of malonyl-CoA. However, genetic manipulation increasing the malonyl-CoA pool failed to increase oil yields significantly, which suggests that there are additional controls in fatty acid synthesis.

The quality of the oil produced can be altered by making single gene changes. The oleic acid content of soybean was increased to 86% by suppression of the enzyme oleoyl desaturase (Thelen and Ohlrogge, 2002). Rapeseed has been engineered to accumulate 58% short-chain lauric acid (12:0) in its oil by expression of a thioesterase from the California bay plant. However, not all gene transfers have been successful and breakdown of the target product can occur in some cases.

One of the possibilities of increasing oil supplies is to introduce new plants, especially those that can be grown on marginal land not suited to food crops. One group of plants is the oilseed shrubs which include castor bean (*Ricinus communis*), *Pongamia pinnata*, *Calophyllum inophyllum* and *Jatropha curcas*. Oil from these plants has been tested for its suitability as biodiesel (Forson *et al.*, 2004; Chapter 7, Table 7.17). Castor oil is best suited as a lubricant, and for many other industrial uses. *J. curcas* has attracted interest as the oil is suitable as a biofuel and the plant can grow in the desert without addition of water and fertilizer (Openshaw, 2000). However, as a commercial crop *J. curcas* has a number of problems. It has to be harvested by hand and contains toxic alkaloids, phorbol esters and curcin which make the meal unsuitable as animal feed. It is perhaps why *J. curcas* is known as black vomit nut, purge nut, physic nut and the extracted oil, hell oil. The other plants also have toxic compounds in their seeds. Castor beans contain ricin, a neurotoxin, along with allergens and *P. pinnata* and *C. inophyllum* both have bitter and poisonous compounds in their seeds. To make these plants commercial crops they will need modification which could involve genetic manipulation. The characteristics which need to be introduced are as follows:

- Dwarf stalks for easy harvesting.
- Suppression of branching to allow for mechanical harvesting.
- Introduction of anti-shattering gene to stop fruit drying and scattering seeds.
- The elimination of the toxic compounds.

The genetic manipulation of microalgae has been demonstrated in a few algal species (Rosenberg *et al.*, 2008). An increase in lipid content in the diatom *Cyclotella cryptica* was attempted by overexpression of ACCase but as found with higher plants it did not have a significant effect on lipid yields. More research is required to determine the control mechanisms of lipid synthesis.

Reduction in lignin

Forest trees are not just sources of building material and paper pulp but are also systems for carbon dioxide sequestration and a source of biofuel. It is proposed that genetic manipulation could increase carbon partition to woody tissues, and increasing cellulose availability for digestion. Lignin content can be changed by modification of gene expression (Groover, 2007). Plant material having less lignin is more digestible when lignocellulose needs to be broken down into sugars for ethanol production. Reduction in lignin content has been achieved in trees in order to reduce the bleaching required when making paper pulp. Changes in lignin content have been carried out in maize, sorghum, poplar and pine (Gressel, 2008). Some concerns have been aired that a reduction in lignin content will weaken plants, causing flattening of crops (lodging), but the evidence for this is limited as many of the crops used now are short stemmed.

Hydrogen production in microalgae

One of the sustainable methods of producing hydrogen is to use anaerobic photosynthetic bacteria which use light and organic acids to produce hydrogen. However, yields of hydrogen are low and are only produced under stress conditions, so that genetic manipulation may be able to increase yields without using stress conditions.

Improvements to the production of ethanol

In addition to improving the production of ethanol by modifications to the process, modern molecular biology offers chances to alter all aspects of ethanol production. In terms of bioethanol genetic modification can be applied both to the biomass sources and the fermenting organisms.

The biomass sources for ethanol are sugars, starches and lignocellulose. Increasing sugar content of sugarcane and sugarbeet plants had reached a limit but recently a sucrose isomerase enzyme has been targeted to sugarcane vacuoles, which converts the glycosidic linkage in sugar to a 1,6-fructoside. This allows the accumulation of 0.5M isomaltose (palatinose) in sugarcane stems in addition to sucrose which increases the overall sugar content (Gressel, 2008). However, the one enzyme capable of degrading isomaltose only degrades it slowly but it is hoped to increase the rate greatly by gene shuffling. A change in starch composition to make its conversion into sugar easier is possible and is under investigation.

The development of some second- and third-generation biofuels depends on the ability to process lignocellulose. If lignocellulose is to be used for ethanol production, it has to be broken down into sugars. The lignin content of wood, straw and grasses reduces the rate of cellulose hydrolysis due to steric hindrance of the cellulolytic enzymes. A reduction in lignin or an increase in cellulose would increase the production of sugars. Plant material with more cellulose and less lignin has been reported where partial silencing of the phenylpropanoid pathway enzymes leading to lignin reduces the lignin content (Morohoshi and Kajita, 2001; Gressel and Zilberstein, 2003). In this way an increase in the digestibility of maize, sorghum, pearl millet, poplar and pine has been achieved. A possible problem with the reduction in lignin is loss of structural strength and subsequent lodging (blowing down in strong winds). The dwarf and semi-dwarf wheat and rice will probably not suffer from this and there appears to be no correlation between lodging and lignin content (Gressel, 2008).

The best and most widely used ethanol-producing microorganism *Saccharomyces cerevisiae* has a rapid growth rate, and a tolerance to ethanol accumulating in the medium. However, *S. cerevisiae* has only a limited substrate range restricted to a few sugars and it is unable to metabolize starch and lignocellulose or the pentose sugar from lignocellulose. There are ethanol-producing bacteria, perhaps the best known is *Zymomonas mobilis*, but these also have a restricted range of sugars that they can ferment. One solution to this dilemma is to use genetic manipulation to introduce the ability to use alternative substrates like starch and xylose into these organisms.

Starch utilization

Saccharomyces cerevisiae has been engineered to express the α-amylase and glucoamylase enzymes from the filamentous fungus *Aspergillus shirousamii* (Shiuya *et al.*, 1992) and has shown a high level of enzyme activity with starch as a substrate. Other constructs have been made with enzymes from *Bacillus subtilis*, *Aspergillus awamori* and mouse pancreatic α-amylase (Birol *et al.*, 1998). Ethanol production levels in starch and glucose-containing media were found to be comparable.

Lignocellulose hydrolysates

The sugars produced by the hydrolysis of lignocellulose are glucose from cellulose, and xylose, arabinose, glucose, mannose and galactose from hemicellulose and galacturonic acid from pectin. *Saccharomyces cerevisiae* can ferment glucose, mannose and galactose but not the other sugars. There are some yeasts which can metabolize xylose, but only a few can ferment xylose to ethanol. One approach was to introduce xylose catabolism into ethanol-producing yeast. Xylose-utilizing yeasts *Pichia stipitis*, *Pachysolen tannophilus* and *Candida shehatae* have very fastidious growth conditions. In *S. cerevisiae*, xylose is reduced to xylitol and xylitol is reduced to xylulose which is slowly metabolized. In the xylose-metabolizing yeasts, xylose is converted into xylulose by xylose reductase and xylitol dehydrogenase. In the reaction NADH and NADP are produced which need to be regenerated and can normally be carried out aerobically, but under anaerobic conditions no electron acceptors are available. A recombinant *S. cerevisiae* has been produced with the complete xylose pathway from *P. stipitis*, but the cells only metabolize xylose aerobically (Fig. 8.31) (Prasad *et al.*, 2007), and under anaerobic conditions xylitol was produced.

Arabinose is also present in lignocellulose hydrolysates which *S. cerevisiae* cannot ferment. Overexpression of all structural genes of a fungal arabinose pathway produced a *S. cerevisiae* strain capable of fermenting arabinose. However, the production was too low to be commercial.

Another approach is to expand the substrate range of ethanol-producing bacteria. One example was to insert the *Z. mobilis* ethanol pathway into microorganisms which can use xylose, such as *Escherichia coli*. This involves the insertion of pyruvate decarboxylase and alcohol dehydrogenase (Fig. 8.32).

Another group of enzymes of interest are the cellulases and genetic manipulation has been used to increase their efficiency and production. To date cellulases have been expressed in *S. cerevisiae*, *Z. mobilis*, *E. coli* and *Klebsiella oxytoca* but with limited success (Chang, 2007).

Energy Balances in Biofuel Production

One of the perceived problems with all biofuels is that if they require too much energy for their production then this makes them uneconomic and unsustainable. The energy input is measured by the net energy value (NEV), the ratio of the energy obtained versus the energy required to generate the fuel. If this is less than one it indicates an unsustainable energy source.

Energy balances for biomass

The net energy balance or NEV for biomass combustion or use in electricity generation is shown in Fig. 8.33. Electricity generated from fossil fuel had an NEV of 0.32 and electricity generation with all forms of biomass had much higher values. This indicates that these are more sustainable sources. Electricity generated from SRC and Miscanthus, either directly or after gasification, had NEV values of around 5. Combustion of willow and the perennial grasses had NEV values of 15 and above. The differences between combustion and electricity generation reflect the losses that occur in electricity generation.

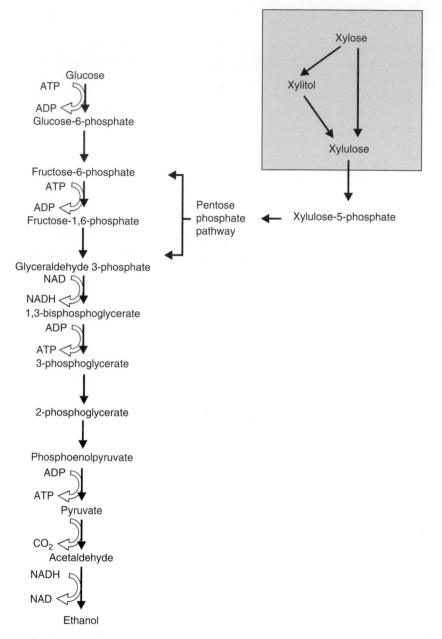

Fig. 8.31. The introduction of xylose-metabolizing enzymes into *Saccharomyces cerevisiae*. The shaded area indicates the introduced enzymes. (From Prasad *et al.*, 2007.)

Ethanol

The production of ethanol has been questioned in terms of energy input (Ho, 1989; Pimentel, 1991, 2004; Niven, 2005). Life-cycle analysis has shown that in some cases the energy input exceeds the energy contained in the fuel, giving a negative NEV. A number of

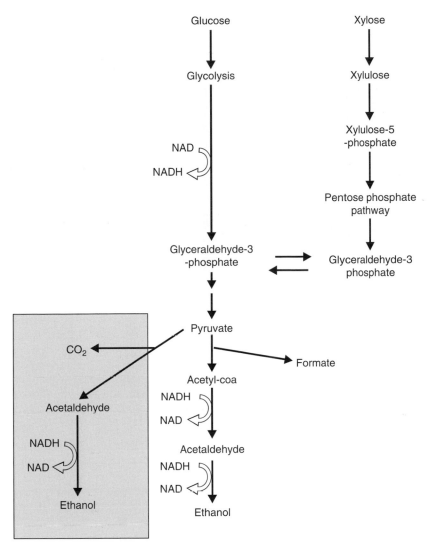

Fig. 8.32. The insertion of the ethanol synthesis pathway from *Z. mobilis* into *E. coli* which can then utilize xylose. (From Chang, 2007.)

studies on the NEV of ethanol production are given in Fig. 8.34 which shows considerable variation in the results. The wide variation has been thought to be due to the different values used to estimate the energy input into crop cultivation, crop yield and ethanol processing. For instance there is considerable difference between Pimentel (1991) and Lorenz and Morris (1995) although both studies have the same inputs. Pimentel includes the energy carried in the farm machinery to give a negative value. Another explanation of the variation over time is that maize yields have risen over the years from 70 bushels/acre (1 acre = 1.609 km²) in 1969 to 135 bushels/acre in 2001. Many of the studies were produced from data collected in the 1980s when yields were lower. The energy content of the co-products can also affect the NEV. Wet milling of maize yields maize oil, gluten meal and gluten feed while dry milling

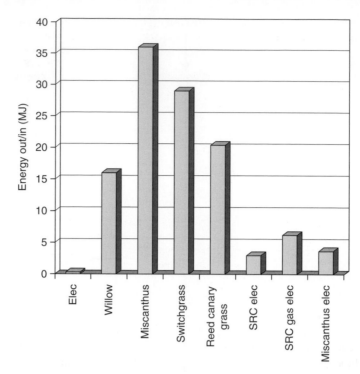

Fig. 8.33. Energy balances for biomass used in combustion and electricity generation. SRC, short rotation coppice. (From Borjesson, 1996; Dubisson and Sinyzoff, 1998; Bullard and Metcalf, 2001; Keoleian and Volk, 2005; Elsayed *et al.*, 2003; Heller *et al.*, 2003; Mortimer *et al.*, 2003; Rowe *et al.*, 2007.)

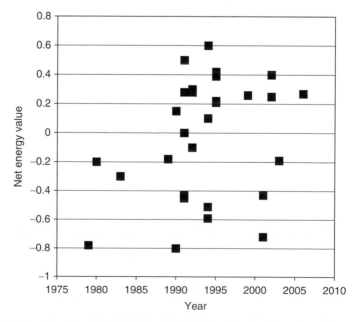

Fig. 8.34. Net energy values for ethanol production from a number of sources.

Table 8.7. The energy required to produce 1000 l of ethanol from maize and sugarcane. (Adapted from Glazer and Nikaido, 1994.)

Input	Maize	Sugarcane
Maize 2,700 kg = 3,259 kcal	3,259	–
Sugarcane 14,000 kg = 1,945 kcal	–	1,945
Transport	325	400
Water	90	70
Steel	228	91
Cement	60	15
Coal	4,617	–
Bagasse	–	7,600
Total input	8,579	10,121
Total output 1,000 l = 5,130 kcal	5,130	5,130
Ratio output/input	0.59	0.51
If the input of the fuels are removed the input is	3,961	2,521
Ratio output/input	1.29	2.03

yields distiller's dried grain with solubles (DDGS) which can be burnt to produce energy. Also energy use in farming has been declining, all of which has made the energy balance for bioethanol positive. An example of the energy input and output for ethanol production from maize and sugarcane is given in Table 8.7. The process using sugarcane uses less energy for cultivation, harvesting and construction than the maize process. Apart from the energy required for cultivation, the major energy input was either coal or bagasse used to generate steam for distillation of ethanol. Adding these to the other inputs means that less energy is obtained than was used, giving a negative NEV. However, if the coal and bagasse are discounted, the NEV is positive. In the case of sugarcane, bagasse is the cellulose waste from sugarcane processing and therefore comes at no cost and burning bagasse makes the process much more energy-positive than the maize process. This particular study does not include any energy values obtained from the by-products and hence may differ from other studies.

Ethanol production is increasing and a United States Department of Agriculture (USDA) report states that bioethanol is energy-efficient and produces 34% more energy than its production uses with a NEV value of 5.28 (5.9 MJ/l) (Shapouri *et al.*, 2002).

Biodiesel

The NEV values for biodiesel produced from rapeseed (RME) using a conventional process and a low nitrogen process, compared with sunflower are given in Fig. 8.35. The values are in general better than ethanol as the process is simpler. The higher NEV value for the low nitrogen shows the effect on energy input that reduction in fertilizer use has. A breakdown of the energy used in the biodiesel process is given in Fig. 8.20a. The largest energy input is esterification followed by nitrogen fertilizer and solvent extraction. The remaining contributions are very small in relation to transesterification and the total energy was 16.3 MJ/kg and in the low nitrogen process the total energy was 7.75 MJ/kg.

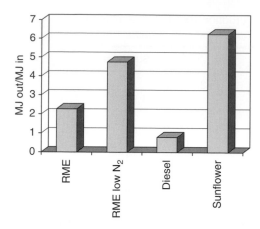

Fig. 8.35. Comparison of the net energy values of diesel and biodiesel produced from rapeseed, conventional and low nitrogen cultivation, and sunflower. (From Mortimer *et al.*, 2003; Powlson *et al.*, 2005; HGCA, 2005.)

Life-cycle Analysis of Biofuels

A life-cycle analysis is often used to demonstrate that one product is environmentally superior to another and to identify stages where a reduction in use of resources and emissions can be achieved. This type of analysis is ideal to determine the impact of biofuels on global warming using a WTW analysis. A WTW analysis combines the fuel production system, well-to-tank (WTT), the energy and GHGs associated with delivering the fuel to the vehicle's tank. The second part is tank-to-wheel (TTW), the energy expended and GHGs produced by the vehicle power train. These are combined to produce the WTW values.

In a recent study, the Joint Research Centre has given the energy use and GHG emissions in a WTW for a number of fuels including ethanol and biodiesel. The study was not a complete life-cycle analysis as it does not include the energy used and GHG emissions produced in building the production facilities and vehicles. Nevertheless the data illustrate the problems with some biofuel sources.

Ethanol

The WTW analysis of energy use and greenhouse emissions for ethanol production is affected by the different methods of producing ethanol (Fig. 8.36). Sugarbeet is the main crop for sugar production in the EU and the energy use can be reduced by using the pulp for animal feed or burning it to produce heat (Malca and Freire, 2006). The lignocellulose route, although not yet commercial, to ethanol using wheat straw and wood also reduces both the energy use and GHG emissions to sugarcane levels. Ethanol from sugarcane has low values for both energy and GHGs as the crop uses little fertilizer and the waste bagasse is used as a fuel. The production of ethanol from sugarbeet has a modest saving of energy and GHGs when the pulp is used as animal feed but this is greatly improved when the pulp is used to generate heat. Sugarcane has considerably lower values than sugarbeet as the waste material bagasse is used to generate heat. The energy and GHGs produced when wheat is used

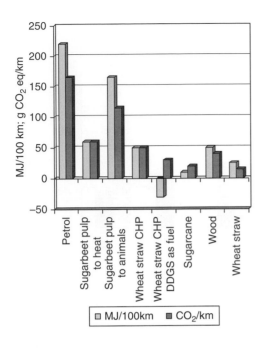

Fig. 8.36. Well-to-wheel (WTW) energy use and carbon dioxide production for ethanol using different substrates and co-product uses such as animal feed or combustion for energy. CHP, combined heat and power; DDGS, distiller's dried grain with solubles. (From JRC, 2007.)

as a source of ethanol can be reduced when the straw is used in combined heat and power systems. The energy can be further reduced when the residue remaining after ethanol production, distiller's dried grain with solubles (DDGS), is used as a fuel.

Biodiesel

The production of biodiesel is less energy-intensive than bioethanol as the process is simpler. The use of a co-product glycerol does affect the energy balance (Fig. 8.37). The rapeseed oil biodiesel reduces both energy and GHG emissions by half compared with diesel, but the fate of the by-product glycerol has little effect. The sunflower oil biodiesel values were a little lower than the rapeseed biodiesel and again the use of the glycerol had little effect on the energy used.

The production of diesel by the FT process is greatly influenced by the starting material. In all three methods of producing diesel, the energy used was greater than fossil diesel and only when wood was used are there gains in greenhouse gas emissions (Fig. 8.38).

Gaseous fuels

The WTW values for gaseous fuels have been compared with CNG and petrol in Fig. 8.39. CNG requires more energy than petrol and diesel but produces about the same quantity of GHGs. When compressed biogas is used, the amount of energy used is greater than the fossil fuels, but biogas produced from liquid, solid manure and municipal solid waste saved considerable amounts of GHGs. Clearly improvements in energy use for biogas are required to make the process sustainable.

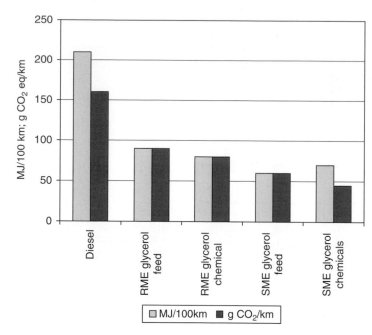

Fig. 8.37. Well-to-wheel (WTW) analysis of energy balance for biodiesel. RME, biodiesel produced from rapeseed with glycerol used as animal feed and chemical synthesis; SME, soybean biodiesel with glycerol used for animal feed and chemical synthesis. (From JRC, 2007.)

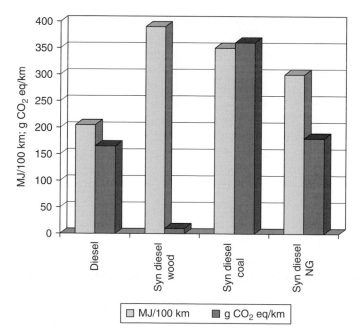

Fig. 8.38. The well-to-wheel (WTW) energy and emissions for the production of syn-diesel using the Fischer-Tropsch process with different starting materials. NG, natural gas. (From JRC, 2007.)

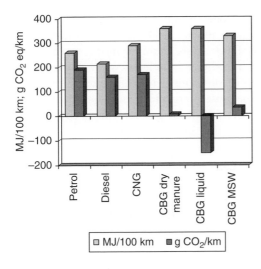

Fig. 8.39. The well-to-wheel (WTW) energy use and greenhouse gas emissions for compressed natural gas (CNG), and compressed biogas (CBG) from dry manure, liquid slurry and municipal solid waste (MSW). (From JRC, 2007.)

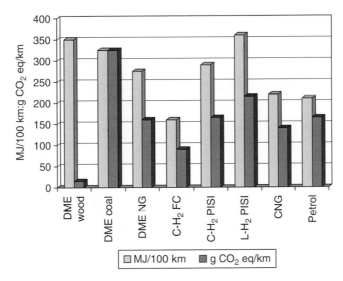

Fig. 8.40. The well-to-wheel (WTW) energy and emissions for gaseous fuels DME and hydrogen. DME produced from wood, coal and natural gas. Hydrogen compressed (C-H$_2$) and liquid (L-H$_2$) in a port injection internal combustion engine and compressed in a fuel cell (C-H$_2$ FC). NG, natural gas; PISI, port injection spark ignition. (From JRC, 2007.)

DME requires more energy for its production than petrol and CNG due to gasification and gas cleaning steps, but if wood is used for its production the emissions are reduced to a very low level (Fig. 8.40). In contrast, DME produced from coal and natural gas has high energy and GHG values. Hydrogen is normally produced from natural gas when used in an internal combustion engine and requires more energy than petrol and produces more emissions, but these are considerably reduced if the hydrogen is used in a fuel cell. Hydrogen stored as a liquid had higher energy and GHG values than hydrogen stored as a compressed gas. The hydrogen values are similar to CNG. The hydrogen figures are very dependent on the method used to generate hydrogen and would be greatly improved if hydrogen was produced either biologically or from sustainable electricity. This can be seen in Fig. 8.41 where the methods of producing hydrogen have been compared. Electrolysis requires more energy than hydrogen production from wood, coal and natural gas, but electricity generated from wind, nuclear and wood reduces the GHG emissions considerably. The use of nuclear-generated electricity does, however, require the most energy. Hydrogen generated from wood also has a greatly reduced emission level. Natural gas reforming to produce hydrogen tends to be more efficient than gasification in terms of energy.

All the data on the WTW have been combined in a scatter figure plotting GHG against energy use (Fig. 8.42). The figure shows considerable variation in the WTW values for the biofuels. The low values for ethanol are from sugarcane production and the low GHG values for hydrogen, biogas and DME are due to using sustainable feedstocks such as wood. The higher GHG emissions are due to the use of fossil fuels such as coal. The ideal fuel would have low values for both GHGs and energy and this can perhaps be achieved by process and feedstock changes. The WTW study can be used to indicate which fuel and feedstock needs improving. Those fuels showing low GHG and energy are ethanol from sugarcane, lignocellulose and biodiesel from plant oils. All those with low GHG emissions have been produced from sustainable materials such as lignocellulose, wastes and plants.

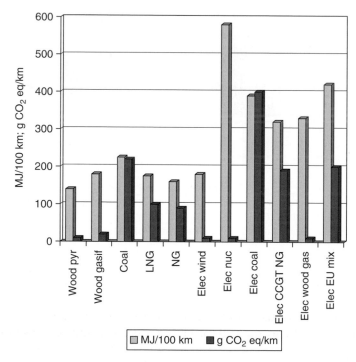

Fig. 8.41. The effect of the method of hydrogen production on energy use and emissions. The direct production from wood by pyrolysis (pyr) and gasification. Direct production from coal, liquid natural gas (LNG) and natural gas (NG). Electrolysis (elec) using wind, nuclear power, coal, combined cycle gas turbine (CCGT) on natural gas, wood gasification and the standard EU electricity mix. (From JRC, 2007.)

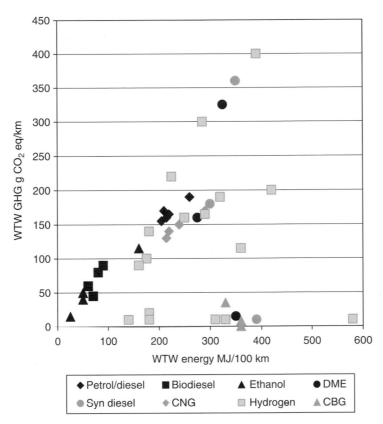

Fig. 8.42. Well-to-wheel (WTW) energy and greenhouse emissions for a variety of biofuels. (From JRC, 2007.)

Infrastructure

Biodiesel is a liquid fuel with a density close to that of diesel, non-toxic, biodegradable, with a high flash point which means that it can be used by the diesel supply infrastructure without any significant changes. In fact it is somewhat safer than diesel. Bioethanol has many of the characteristics of petrol except that it is hydroscopic. The ability to accumulate water means that bioethanol cannot be transferred by pipeline because the water content causes corrosion (rust). This means that it has to be transferred by tanker and this is proving to be a problem in the USA where the transport provisions have not kept pace with ethanol production.

The gaseous fuels methane and hydrogen pose different problems from the liquid fuels. Methane is essentially the same as natural gas and could be transported through the natural gas infrastructure. Hydrogen once compressed can be run through pipelines with a loss of 0.77% per 100 km but with a low density a leak will disperse rapidly (Hammerschlag and Mazza, 2005).

If hydrogen is produced in situ by the electrolysis of water the efficiency is 74% and compression is 88% efficient, which means that only 65% of the energy will be delivered as hydrogen. Hammerschlag and Mazza (2005) quote: 'In virtually any conceivable

Table 8.8. Alternative fuel stations in the UK. (Adapted from Energy Saving Trust, 2007.)

Fuel	December 2005	August 2007
Liquid petroleum gas (LPG)	1192	1490
Compressed natural gas	25	18
Recharging points	13	20
Biodiesel	106	151
Ethanol	0	14
Hydrogen	0	0

arrangement for supplying renewable or carbon-neutral energy to electric customers, delivering the electricity directly is more efficient than manufacturing hydrogen.' Table 8.8 gives the fuel stations capable of supplying a range of alternative transport fuels including biodiesel and bioethanol for the years 2005 and 2007. Worldwide there are 140 stations supplying hydrogen but none for private cars in the UK.

Conclusions

Biofuels should not be considered in isolation as alternatives to fossil fuels but as a part of a drive towards the production of sustainable products normally produced from crude oil. Biorefineries which can produce a range of products including biofuels from renewable resources should be developed.

Liquid fuels of all types will continue to be used for some time because of the difficulty of supplying alternative biofuels, and the existing extensive infrastructure will continue to be used. The use of gaseous fuels such as hydrogen, CNG and LNG require extensive and costly modification to vehicles and in the case of hydrogen a completely new infrastructure. Thus, these are long-term solutions. Fuel cells may replace the internal combustion engine but these are still under development and will require modification according to the supply infrastructure, depending on the fuel used in the fuel cells. Bioethanol and biodiesel are fuels that can be used now in present vehicles and infrastructure. The main restrictions on these fuels are insufficient supply and their cost, which could be improved by the tax structure. FT diesel, FT petrol and DME are in a developmental stage, as technical advances are required to make their production economic. In the longer term, hybrid or electric cars may be the best option for short-distance travel and city use provided the cars are charged using renewable-generated electricity. Diesel and biodiesel will probably be retained for heavy transport; trains and cars will also use more diesel as this has better fuel consumption. It is air transport that has not been addressed, probably because fuel is cheap. Biodiesel can replace kerosene and successful tests have been carried out with turboprop engines.

References

Abigor, R., Uadia, P., Foglia, T., Haas, M., Jones, K., Okpefa, E., Obibuzor, J. and Bafor, M. (2000) Lipase-catalysed production of biodiesel fuel from some Nigerian lauric oils. *Biochemical Society Transactions* 28, 979–981.

Acien Fernandez, F.G., Hall, D.O., Guerrero, E.C., Rao, K.K. and Molina Grima, E. (2003) Outdoor production of *Phaeodactylum tricornutum* biomass in a helical reactor. *Journal of Biotechnology* 103, 137–152.

Agarwal, A.K. (2007) Biofuels (alcohols and biodiesel) applications as fuel for internal combustion engines. *Progress in Energy and Combustion Science* 33, 233–271.

Ahman, I.M., Kazachkova, N.I., Kammert, I.M., Hagberg, P.A., Dayteg, C.I., Eklund, G.M., Meijer, L.J.O. and Ekbom, B. (2006) Characterisation of transgenic oilseed rape expressing pea lectin in anthers for improving resistance to pollen beetle. *Euphytica* 151, 321–330.

Alcantara, R., Amores, J., Canoira, L., Fidalo, E., Franco, M.J. and Navarro, A. (2000) Catalytic production of biodiesel from soy-bean oil, used frying oil and tallow. *Biomass Bioenergy* 18, 515–527.

Aleklett, K. (2005) Impact on oil depletion to 2020 for renewable fuels and developments. Available at http://www.peakoil.net

Al-Alawi, A., van de Voort, F.R., Sedman, J. and Ghetler, A. (2006) Automated FTIR analysis of free fatty acids or moisture in edible oils. *Journal of the Association for Laboratory Automation* 11, 23–29.

Altiparmak, D., Keskin, A., Koca, A. and Guru, M. (2007) Alternative fuel properties of tall oil fatty acid methyl-ester fuel blends. *Bioresource Technology* 98, 241–246.

Al-Widyan, M.I., Tashtoush, G. and Abu-Qudais, M. (2002) Utilization of ethyl ester of waste vegetable oils as fuel in diesel engines. *Fuel Processing Technology* 76, 91–103.

Antolin, G., Tinaut, F.V., Briceno, Y., Castano, V., Perez, C. and Ramirez, A.I. (2002) Optimisation of biodiesel production by sunflower oil transformation. *Bioresource Technology* 83, 111–114.

Antoni, D., Zverlov, V.V. and Schwarz, W.H. (2007) Biofuels from microbes. *Applied Microbiology and Biotechnology* 77, 23–35.

Arechederra, R.L., Treu, B.L. and Minteer, S.D. (2007) Development of glycerol/O_2 biofuel cell. *Journal of Power Sources* 173, 156–161.

Aresta, M., Dibenedetto, A., Carone, M., Colonna, T. and Fragale, C. (2005) Production of biodiesel from macroalgae by supercritical CO_2 extraction and thermochemical liquefaction. *Environmental Chemistry Letters* 3, 136–139.

Ashby, R.D., Solaiman, D.K.Y. and Fogia, T.A. (2004) Bacterial poly(hydroxyalkanoate) polymer production from the biodiesel co-product stream. *Journal of Polymers and the Environment* 12, 105–112.

Azar, C. (2005) Emerging scarcities: Bioenergy–food competition in a carbon constrained world. In: Simpson, R.D., Toman, M.A. and Ayers, R.U. (eds) *Scarcity and Growth Revisited, Resources for the Future*. RFF, Washington, DC.

Balat, M., Balat, H. and Oz, C. (2007) Progress in bioethanol processing. *Progress in Energy & Combustion Science* 34, 551–573.

Ballesteros, M., Oliva, J.M., Negro, M.J., Manzanares, P. and Ballesteros, I. (2004) Ethanol from lignocellulose materials by a simultaneous saccharification and fermentation process (SSF) with *Kluyveromyces marxianus* CECT 10875. *Process Biochemistry* 39, 1843–1848.

Ban, K., Kaieda, M., Matsumoyo, T., Kondo, A. and Fukuda, H. (2001) Whole cell biocatalyst for biodiesel fuel production utilizing *Rhizopus oryzae* cells immobilized within biomass support particles. *Biochemical Engineering Journal* 8, 39–43.

Banerjee, A., Sharma, R., Chisti, Y. and Banerjee, U.C. (2002) *Botryococcus braunii*: a renewable source of hydrocarbons and other chemicals. *Critical Reviews in Biotechnology* 22, 245–279.

Bastianoni, S., Coppola, F., Tiezzi, E., Colacevich, A., Borghini, F. and Focardi, S. (2007) Biofuel potential production from the Orbetello lagoon macroalgae: a comparison with sunflower feedstock. *Biomass Bioenergy* 32, 619–628.

Becker, E.W. (2007) Micro-algae as a source of protein. *Biotechnology Advances* 25, 207–210.

Belarbi, E.H., Molina, E. and Chisti, Y. (2000) A process for high yield and scaleable recovery of high purity eicosapentaenoic acid esters from microalgae and fish oil. *Enzyme and Microbial Technology* 26, 516–529.

Benemann, J.R. (1997) CO_2 mitigation with microalgal systems. *Energy Conversion & Management* 38, S475–S479.

Bentley, R.W. (2002) Global oil & gas depletion: an overview. *Energy Policy* 30, 189–205.

Bentley, R.W., Mannan, S.A. and Wheeler, S.J. (2007) Assessing the date of global oil peak: the need to use 2P reserves. *Energy Policy* 35, 6364–6382.

Bernardo, A., Howard-Hildige, R., O'Connell, A., Nichol, R., Ryan, J., Rice, B., Roche, E. and Leahy, J.J. (2003) Camelina oil as a fuel for diesel transport engines. *Industrial Crops & Products* 17, 191–197.

Berndes, G., Hoogwijk, M. and van der Broek, R. (2003) The contribution of biomass in the future global energy supply: a review of 17 studies. *Biomass Bioenergy* 25, 1–28.

Bioengineering Resources Inc. (BRI) (2007) Available at http://www.brienergy.com

Birol, G., Onsan, I., Kirdar, B. and Oliver, S.G. (1998) Ethanol production and fermentation characteristics of recombinant *Saccharomyces cerevisiae* strains grown on starch. *Enzyme Microbial Technology* 22, 672–677.

Bode, S. (2006) Long-term greenhouse gas emission reductions – what's possible, what's necessary? *Energy Policy* 34, 971–974.

Bomb, C., McCormick, K., Deuwaarder, E. and Kaberger, T. (2007) Biofuels for transport in Europe: lessons from Germany and the UK. *Energy Policy* 35, 2256–2267.

Bondioli, P., Gasparoli, A., Bella, L.D., Tagliabue, S. and Toso, G. (2003) Biodiesel stability under commercial storage conditions over one year. *European Journal of Lipid Science and Technology* 105, 735–741.

Bondioli, P., Gasparoli, A., Bella, L.D., Tagiabue, S., Lacoste, F. and Lagardere, L. (2004) The prediction of biodiesel storage stability: proposal for a quick test. *European Journal of Lipid Science and Technology* 106, 822–830.

Borjesson, P. (1996) Energy analysis of biomass production and transportation. *Biomass Bioenergy* 11, 305–318.

Borowitzka, M.A. (1999) Commercial production of microalgae: ponds, tanks, tubes and fermenters. *Journal of Biotechnology* 70, 313–321.

Bouaid, A., Diaz, Y., Martinez, M. and Aracil, J. (2005) Pilot plant studies of biodiesel production using *Brassica carinata* as raw material. *Catalysis Today* 106, 193–196.

Boyle, G. (1996) *Renewable Energy: Power for a Sustainable Future*. Oxford University Press, Oxford.

Brammer, J.G., Laner, M. and Bridgewater, A.V. (2006) Opportunities for biomass-derived 'bio-oil' in European heat and power markets. *Energy Policy* 34, 2871–2880.

British Petroleum (BP) (2005) Statistical review. Available at http://www.bp.com

Bullard, M. and Metcalfe, P. (2001) Estimating the energy requirements and CO_2 emission from production of the perennial grasses Miscanthus, switchgrass and reed canary grass. ADAS Consulting Ltd. URN 01/797. Dti.

Business Enterprise & Regulatory Reform Department (BERR) (2007) UK energy in brief July 2007. Available at http://www.berr.gov.uk

Canakci, M. (2007) The potential of restaurant waste lipids as biodiesel feedstocks. *Bioresource Technology* 98, 183–190.

Canakci, M. and van Gerpen, J. (1999) Biodiesel production via acid catalysis. *Transactions of the ASAE* 42, 1203–1210.

Cannell, M.G.R. (2003) Carbon sequestration and biomass energy offset: theoretical, potential and achievable capacities globally, in Europe and in the UK. *Biomass Bioenergy* 24, 97–116.

Canoira, L., Alcantara, R., Garcia-Martinez, M.J. and Carrasco, J. (2006) Biodiesel from Jojoba oil-wax: transesterification with methanol and properties as a fuel. *Biomass Bioenergy* 30, 76–81.

Cao, W., Han, H. and Zhang, J. (2005) Preparation of biodiesel from soybean oil using supercritical methanol and co-solvent. *Fuel* 84, 347–351.

Cardona, C.A. and Sanchez, O.J. (2007) Fuel ethanol production: process design trends and integration opportunities. *Bioresource Technology* 98, 2415–2457.

Carraretto, C., Macor, A., Mirandola, A., Stoppato, A. and Tonon, S. (2004) Biodiesel as alternative fuel: experimental analysis and energetic evaluations. *Energy* 29, 2195–2211.

Cedigaz (2004) Available at http://www.cedigaz.org

Chae, S.R., Hwang, E.J. and Shin, H.S. (2006) Single cell protein production of *Euglena gracilis* and carbon dioxide fixation in an innovative photo-bioreactor. *Bioresource Technology* 97, 322–329.

Champagne, P. (2007) Feasibility of producing bio-ethanol from waste residues: a Canadian perspective. *Resources Conservation & Recycling* 50, 211–230.

Chang, M.C.Y. (2007) Harnessing energy from plant biomass. *Current Opinion in Chemical Biology* 11, 1–8.

Charlier, R.H. (2007) Forty candles for the Rance River TPP tides provide renewable and sustainable power generation. *Renewable & Sustainable Energy Reviews* 11, 2032–2057.

Cheng, L., Zhang, L., Chen, H. and Gao, C. (2006) Carbon dioxide removal from air by microalgae cultured in a membrane-photobioreactor. *Separation Purification Technology* 50, 324–329.

Chiaramonti, D., Bonini, M., Fratini, E., Tondi, G., Gartner, K., Bridgewater, A.V., Grimm, H.P., Soldaini, I., Webster, A. and Baglioni, P. (2003) Development of emulsions from biomass pyrolysis liquid and diesel and their uses in engines – part 1: emulsion production. *Biomass Bioenergy* 25, 85–99.

Chisti, Y. (2007) Biodiesel from microalgae. *Biotechnology Advances* 25, 294–306.

Chiu, C.W., Schumacher, L.G. and Suppes, G.J. (2004) Impact of cold flow improvers on soybean biodiesel blend. *Biomass Bioenergy* 27, 485–491.

Chum, H.L. and Overend, R.P. (2001) Biomass and renewable fuels. *Fuel Processing Technology* 71, 187–195.

Clark, W.W. and Rifkin, J. (2006) A green hydrogen economy. *Energy Policy* 34, 2630–2639.

Clemens, J., Trimborn, M., Weiland, P. and Amon, B. (2006) Mitigation of greenhouse gas emissions by anaerobic digestion of cattle slurry. *Agriculture Ecosystems & Environment* 112, 171–177.

Cocco, D., Tola, V. and Cau, G. (2006) Performance evaluation of chemically recuperated gas turbine (CRGT) power plants fuelled by di-methyl-ether (DME). *Energy* 31, 1446–1458.

Cockroft, J. and Kelly, N. (2006) A comparative assessment of future heat and power sources for the UK domestic sector. *Energy Conversion & Management* 47, 2349–2360.

Cook, J. and Beyea, J. (2000) Bioenergy in the United States: progress and possibilities. *Biomass Bioenergy* 18, 441–455.

Coontz, R. and Hanson, B. (2004) Not so simple. *Science* 305, 957–976.

Cooper, J. (2006) Science & Technology Review. Lawrence Livermore National Laboratory. Available at http://www.llnl.gov

Crabbe, E., Nolasco-Hipolito, C., Koybayashi, G., Sonomoto, K. and Ishizaki, A. (2001) Biodiesel production from crude palm oil and evaluation of butanol extraction and fuel properties. *Process Biochemistry* 37, 65–71.

Crookes, R.J. (2006) Comparative bio-fuel performance in internal combustion engines. *Biomass Bioenergy* 30, 461–468.

Csordas, A. and Wang, J.K. (2004) An integrated photobioreactor and foam fractionation unit for the growth and harvest of *Chaetoceros* sp. in open systems. *Aquacultural Engineering* 30, 15–30.

Cvengros, J. and Cvengrosova, Z. (2004) Used frying oil and fats and their utilization in the production of methyl esters of higher fatty acids. *Biomass Bioenergy* 27, 173–181.

Dai, D., Hu, Z., Pu, G., Li, H. and Wang, C. (2006) Energy efficiency and potentials of cassava fuel ethanol in Guangxi region of China. *Energy Conversion & Management* 47, 1686–1699.

Das, D. and Veziroglu, T.N. (2001) Hydrogen production by biological processes: a survey of literature. *International Journal of Hydrogen Energy* 26, 13–28.

Defra (2007) Environmental statistics. Available at http://www.defra.gov.uk

Defra (2008) Environmental statistics. Available at http://www.defra.gov.uk

Del Campo, J.A., Moreno, J., Rodriguez, H., Vargas, M.A., Rivas, J. and Guerrero, M.G. (2000) Carotenoid content of chloropycean microalgae: factors determining lutein accumulation in *Muriellopsis* sp. (Chlorophyta). *Journal of Biotechnology* 76, 51–59.

Dell, R.M. and Rand, D.A.J. (2001) Energy storage – a key technology for global energy sustainability. *Journal of Power Sources* 100, 2–17.

Demirbas, A. (2006a) Biodiesel production via non-catalytic SCF method and biodiesel fuel characteristics. *Energy Conversion & Management* 47, 2271–2282.

Demirbas, A. (2006b) Global biofuel strategies. *Energy Education Science and Technology* 17, 27–63.

Demirbas, A. (2008) Biofuels sources, biofuel policy, biofuel economy and global biofuel projections. *Energy Conversion & Management* 49, 2106–2116.

De Morais, M.G. and Costa, J.A.V. (2007a) Biofixation of carbon dioxide by *Spirulina* sp. and *Scenedesmus obliquus* cultivated in a three-stage serial tubular photobioreactor. *Journal of Biotechnology* 129, 439–445.

De Morais, M.G. and Costa, J.A.V. (2007b) Carbon dioxide fixation by *Chlorella kessleri, C. vulgaris, Scenedesmus obliquus* and *Spirulina* sp. cultivated in flasks and vertical tubular photobioreactors. *Biotechnology Letters* 29, 1349–1352.

De Paula, G.O. and Cavalcanti, R.N. (2000) Ethics: essence for sustainability. *Journal of Cleaner Production* 8, 109–117.

De Wit, M.P. and Faaij, A.P.C. (2007) Impact of hydrogen onboard storage technologies on the performance of hydrogen fuelled vehicles: a techno-economic well-to-wheel assessment. *International Journal of Hydrogen Energy* 32, 4859–4870.

Department of Trade and Industry (Dti) (2003) *Our Energy Future – Creating a Low Carbon Economy*. The Stationery Office, London.

Department of Trade and Industry (Dti) (2006a) UK energy in brief. July 2006. Available at http://www.dti.gov.uk

Department of Trade and Industry (Dti) (2006b) *The Energy Challenge. Energy Review Report 2006*. The Stationery Office, Norwich.

Dessus, B., Devin, B. and Pharabed, F. (1992) World potential of renewable energies: actually accessible in the nineties and environmental impact analysis. La Huille Blanche No. 1.

Di Serio, M., Tesser, R., Dimiccoli, M., Cammarota, F., Nastasi, M. and Sntacesaria, E. (2005) Synthesis of biodiesel via homogeneous Lewis acid catalyst. *Journal of Molecular Catalysis A: Chemical* 239, 111–115.

Dismukes, G.C., Carrieri, D., Bennette, N., Ananyev, G.M. and Posewitz, M.C. (2008) Aquatic phototrophs: efficient alternatives to land-based crops for biofuels. *Current Opinion in Biotechnology* 19, 235–240.

Dorado, M.P., Ballesteros, E., Arnal, J.M., Gomez, J. and Lopez, F.J. (2003) Exhaust emissions from a diesel engine fueled with transesterified waste olive oil. *Fuel* 82, 1311–1315.

Dote, Y., Sawayama, S., Inoue, S., Minowa, T. and Yokoyama, S. (1993) Recovery of liquid fuel from hydrocarbon-rich microalgae by thermochemical liquefaction. *Fuel* 73, 1855–1857.

Doucha, J. and Livansky, K. (2006) Productivity, CO_2/O_2 exchange and hydaulics in outdoor open high density microalgal (*Chlorella* sp) photobioreactors operated in a Middle and Southern European climate. *Journal of Applied Phycology* 18, 811–826.

Du, D., Sato, M., Mori, M. and Park, E.Y. (2006) Repeated production of fatty acid methyl esters with activated bleaching earth in solvent-free system. *Process Biochemistry* 41, 1849–1853.

Du, Z., Li, H. and Gu, T. (2007) A state of the art review on microbial fuel cells: a promising technology for wastewater treatment and bioenergy. *Biotechnology Advances* 25, 464–482.

Dubuisson, X. and Sintzoff, I. (1998) Energy and CO_2 balances in different power generation routes using wood fuel from short rotation coppice. *Biomass Bioenergy* 15, 379–390.

Dunn, R.O. (2005) Effect of antioxidants on the oxidative stability of methyl soyate (biodiesel). *Fuel Processing Technology* 86, 1071–1085.

Du Plessis, L.M., De Villiers, J.B.M. and van der Walt, W.H. (1985) Stability studies on methyl and ethyl fatty acid esters of sunflower oil. *Journal of American Oil Chemists' Society* 62, 748–752.

Durre, P. (1998) New insights and novel developments in clostridial acetone/butanol/isopropanol fermentation. *Applied Microbiology and Biotechnology* 49, 639–648.

Ebbinghaus, A. and Weisen, P. (2001) Aircraft fuels and their effect upon engine emissions. *Air & Space Europe* 3, 101–103.

Edinger, R. and Kaul, S. (2000) Humankind's detour toward sustainability: past, present, and future of renewable energies and electric power generation. *Renewable & Sustainable Energy Reviews* 2, 295–313.

Elias, J.A.L., Voltina, D., Ortega, C.O.C., Rodriguez, B.B.R., Gaxiola, L.M.S., Esquivel, B.C. and Nieves, M. (2003) Mass production of microalgae in six commercial shrimp hatcheries of the Mexican northwest. *Aquacultural Engineering* 29, 155–164.

Elsayed, M.A., Matthews, R. and Mortimer, N.D. (2003) Carbon and energy balances for a range of biofuel crops. URN 03/086 Dti.

Enciner, J.M., Gonzalez, J.F., Rodriguez, J.J. and Tajedor, A. (2002) Biodiesel fuel from vegetable oils: transesterification of *Cynara cardunculus* L. oils with ethanol. *Energy Fuels* 16, 443–450.

Energy Information Administration (EIA) (2008) *Annual Energy Outlook 2008*. U.S. Department of Energy, Washington, DC. Available at http://www.eia.gov

Energy Saving Trust (2007) Available at http://www.est.org.uk

Environmental Protection Agency (EPA) (2002) A comprehensive analysis of biodiesel impacts on exhaust emissions. EPA420-P-02–001. Available at http://www.epa.gov

Eriksen, N.T., Poulsen, B.R. and Iversen, J.J.L. (1998) Dual sparging laboratory-scale photobioreactor for continuous production of microalgae. *Journal of Applied Phycology* 10, 377–382.

European Biodiesel Board (2007) Available at http://www.ebb-eu.org

European Biomass Industry Association (Eubia) (2005) Available at http://www.eubia.org

Evans, J. (1999) Call out the reserves. *Chemistry in Britain* 35(8), 38–41.

Evans, J. (2000) Power to the people. *Chemistry in Britain*, August 2000, 30–33.

Faaij, A.P.C. (2006) Bio-energy in Europe: changing technology choices. *Energy Policy* 34, 322–342.

Farrell, A.E., Plevin, R.J., Turner, B.T., Jones, A.D., O'Hare, M. and Kammen, D.M. (2006) Ethanol can contribute to energy and environmental goals. *Science* 311, 506–508.

Fedorov, A.S., Kosourov, S., Ghirardi, M.L. and Seibert, M. (2005) Continuous H_2 photoproduction by *Chlamydomonas reinhartii* using a novel two-stage sulphate-limited chemostat system. *Applied Biochemistry and Biotechnology* 24, 403–412.

Fernando, S., Adhikari, S., Kota, K. and Bandi, R. (2007) Glycerol based automotive fuels from future biorefineries. *Fuel* 86, 2806–2809.

Feron, P.H.M. and Jansen, A.E. (1995) Capture of carbon dioxide using membrane gas absorption and reuse in the horticultural industry. *Energy Conversion Management* 36, 411–414.

Fischer, G. and Schrattenholzer, L. (2001) Global bioenergy potentials through 2050. *Biomass Bioenergy* 20, 151–159.

Fleck-Schneider, P., Lehr, F. and Posten, C. (2007) Modelling of growth and product formation of *Porphyridium purpureum*. *Journal of Biotechnology* 132, 134–141.

Foidl, N., Foidl, G., Sanchez, M., Mittlebach, M. and Hackle, S. (1996) *Jatropha curcas* L as a source for the production of biofuel in Nicaragua. *Bioresource Technology* 58, 77–82.

Forson, F.K., Oduro, E.K. and Hammond-Donkoh, E. (2004) Performance of jatropha oil blends in a diesel engine. *Renewable Energy* 29, 1135–1145.

Freedman, B., Butterfield, R.O. and Pryde, E.H. (1986) Transesterification kinetics of soybean oil. *Journal of the American Oil Chemists' Society* 63, 1375–1380.

Frohlich, A. and Rice, B. (2005) Evaluation of *Camelina sativa* oil as a feedstock for biodiesel production. *Industrial Crops & Products* 21, 25–31.

Fukuda, H., Kondo, A. and Noda, H. (2001) Biodiesel fuel production by transesterification of oils. *Journal of Bioscience and Bioengineering* 92, 405–416.

Fulkerson, W., Judkins, R.R. and Sanghvi, M.K. (1990) Energy from fossil fuels. *Scientific American*, September, 83–89.

Furuta, S., Matsuhashi, H. and Arata, K. (2004) Biodiesel fuel production with solid superacid catalysis in fixed bed reactor under atmospheric pressure. *Catalysis Communications* 5, 721–723.

Furuta, S., Matsuhashi, H. and Arta, K. (2006) Biodiesel fuel production with solid amorphous-zirconia catalysis in fixed bed reactor. *Biomass Bioenergy* 30, 870–873.

Ghadge, S.V. and Rahman, H. (2005) Biodiesel production from mahua (*Madhuca indica*) oil having high free fatty acids. *Biomass Bioenergy* 28, 601–605.

Gielen, D. and Unander, F. (2005) Alternative fuels: an energy technology perspective. International Energy Agency/Energy Technology Office. Available at http://www.iea.org

Glasby, G.P. (2003) Potential impact on climate of the exploitation of methane hydrate deposits offshore. *Marine & Petroleum Geology* 20, 163–175.

Glasby, G.P. (2006) Drastic reductions in utilizable fossil fuel reserves: an environmental imperative. *Environment, Development & Sustainability* 8, 197–215.

Glazer, A.N. and Nikaido, H. (1994) *Microbial Biotechnology*. Freeman & Co, New York.

Gouveia, L., Nobre, B.P., Marcelo, F.M., Mrejen, S., Cardoso, M.T., Palavra, A.F. and Mendes, R.L. (2007) Functional food oil coloured by pigments extracted from microalgae with supercritical CO_2. *Food Chemistry* 101, 717–723.

Gowdy, J. and Julia, R. (2007) Technology and petroleum exhaustion: evidence from two mega-oilfields. *Energy* 32, 1448–1454.

Graboski, N.S. and McCormick, R.L. (1998) Combustion of fat and vegetable oil derived fuels in diesel engines. *Progress in Energy and Combustion Science* 24, 125–164.

Green, C., Baksi, S. and Dilmaghani, M. (2007) Challenges to a climate stabilizing energy future. *Energy Policy* 35, 616–626.

Greene, D.L., Hoson, J.L. and Li, J. (2006) Have we run out of oil yet? Oil peaking analysis from an optimist's perspective. *Energy Policy* 34, 515–531.

Greer, D. (2005) Creating cellulosic ethanol: spinning straw into fuel. *BioCycle* 46, 61–65.

Gressel, J. (2008) Transgenics are imperative for biofuel crops. *Plant Science* 74, 246–263.

Gressel, J. and Zilberstein, A. (2003) Let them eat (GM) straw. *Trends Biotechnology* 21, 525–530.

Grimston, M.C., Karakoussis, V., Fouquet, R., van der Vorst, R., Pearson, P. and Leach, M. (2001) The European and global potential of carbon dioxide sequestration in tacking climate change. *Climate Policy* 1, 155–171.

Grobbelaar, J.U. (1994) Turbulence in mass algal cultures and the role of light/dark fluctuations. *Journal of Applied Phycology* 6, 331–335.

Gronkvist, S., Bryngelsson, M. and Westermark, M. (2006) Oxygen efficiency with regard to carbon capture. *Energy* 31, 3220–3226.

Groover, A.T. (2007) Will genomics guide a greener forest biotech? *Trends in Plant Science* 12, 234–238.

Grubb, M. (2001) Who's afraid of atmospheric stabilization? Making the link between energy resources and climate change. *Energy Policy* 29, 837–845.

Gryglewicz, S. (1999) Rapeseed oil methyl esters preparation using heterogeneous catalysts. *Bioresource Technology* 70, 249–253.

Guseo, R., Valle, A.D. and Guidolin, M. (2007) World oil depletion models: price effects compared with strategic or technological interventions. *Technological Forecasting and Social Change* 74, 452–469.

Gustavsson, L., Borjesson, P., Johansson, B. and Svenningsson, P. (1995) Reducing CO_2 emissions by substituting biomass for fossil fuels. *Energy* 20, 1097–1113.

Haas, M.J. (2005) Improving the economics of biodiesel production through the use of low value lipids as feedstocks: vegetable oil soapstock. *Fuel Processing Technology* 86, 1087–1096.

Hamelinck, C.N. and Faaij, A.P.C. (2006) Outlook for advanced biofuels. *Energy Policy* 34, 3268–3283.

Hamelinck, C.N., Faaij, A.P.C., Uil, H.D. and Boerrigter, H. (2004) Production of FT transportation fuels from biomass: technical options, process analysis and optimization, and development potential. *Energy* 29, 1743–1771.

Hamelinck, C.N., Hooijdonk, G.V. and Faaij, A.P.C. (2005) Ethanol from lignocellulosic biomass: techno-economic performance in short-, middle-, and long-term. *Biomass Bioenergy* 28, 384–410.

Hammerschlag, R. and Mazza, P. (2005) Questioning hydrogen. *Energy Policy* 33, 2039–2043.

Hammond, G.P. (1998) Alternative energy strategies for the United Kingdom revisited. *Technological Forecasting & Social Change* 59, 131–151.

Happe, T., Hemschemeier, A., Winkler, M. and Kaminski, A. (2002) Hydrogenases in green algae: do they save the algae's life and solve our energy problems? *Trends in Plant Science* 7, 246–250.

Heffel, J.W. (2003) NOx emission reduction in a hydrogen fueled internal combustion engine at 3000 rpm using exhaust gas recirculation. *International Journal of Hydrogen Energy* 28, 1285–1292.

Hein, K.R.G. (2005) Future energy supply in Europe: challenge and chances. *Fuel* 84, 1189–1194.

Heller, M.C., Keoleian, G.A. and Volk, T.A. (2003) Life cycle assessment of a willow bioenergy cropping system. *Biomass Bioenergy* 25, 147–165.

Henstra, A.M., Sipma, J., Rinzema, A. and Stams, A.F.M. (2007) Microbiology of synthesis gas fermentation for biofuel production. *Current Opinion in Biotechnology* 18, 200–2006.

Herbert, G.M.J., Iniyan, S., Sreevalsan, E. and Rajapandian, S. (2007) A review of wind energy technologies. *Renewable & Sustainable Energy Reviews* 11, 1117–1145.

Hill, J., Nelson, E., Tilman, D., Polansky, S. and Tiffany, D. (2006) Environmental, economic and energetic costs and benefits of biodiesel and ethanol biofuels. *Proceedings of the National Academy of Sciences of the USA* 103, 11206–11210.

Ho, S.P. (1989) Global warming impact of ethanol versus gasoline. 1989 National Conference, Clean Air Issues and America's Motor Fuel Business, Washington DC.

Hoffert, M.I., Caldeira, K., Benford, G., Criswell, D.R., Green, C., Herzog, H., Jain, A.K., Kheshgi, H.S., Lackner, K.S., Lewis, J.S., Lightfoot, H.D., Manheimer, W., Mankins, J.C., Maul, M.E., Perkins, J., Schlesinger, M.E., Volk, T. and Wigley, T.M.L. (2002) Advanced technology paths to global climate stability: energy for a greenhouse planet. *Science* 298, 981–987.

Home-Grown Cereals Authority (HGCA) (2005) Environmental impact of cereals and oilseed rape for food and biofuels in the UK. Available at http://www.hgca.com

Hoogwijk, M., Faaij, A., van der Broek, R., Berndes, G., Gielen, D. and Turkenburg, W. (2003) Exploration of the range of the global potential of biomass for energy. *Biomass Bioenergy* 25, 119–133.

Hosamni, K.M., Ganjihal, S.S. and Chavadi, D.V. (2004) *Alternanthera triandra* seed oil: a moderate source of ricinoleic acid and its possible industrial utilisation. *Industrial Crops & Products* 19, 133–136.

Hoser, R.A. and O'Kuru, R.H. (2006) Transesterified milkweed (*Asclepias*) seed oil as a bio-diesel fuel. *Fuel* 85, 2106–2110.

Houghton, J.J., Jenkins, G.J. and Ephraums, J.J. (1990) *Climate Change*. Cambridge University Press, Cambridge.

Hu, J., Du, Z., Li, C. and Min, E. (2005) Study on the lubrication properties of biodiesel as fuel lubricity enhancers. *Fuel* 84, 1601–1606.

Huang, Z.H., Ren, Y., Jiang, D.M., Liu, L.X., Zeng, K., Liu, B. and Wang, X.B. (2006) Combustion and emission characteristics of a compression ignition engine fuelled with Diesel-dimethoxy methane blends. *Energy Conversion & Management* 47, 1402–1415.

Hutchinson, J.J., Campbell, C.A. and Desjardins, R.L. (2007) Some perspectives on carbon sequestration in agriculture. *Agricultural & Forest Meteorology* 142, 288–302.

Ikura, M., Stanciulescu, M. and Hogan, E. (2003) Emulsification of pyrolysis derived bio-oil in diesel fuel. *Biomass Bioenergy* 24, 221–232.

Illman, A.M., Scragg, A.H. and Shales, S.W. (2000) Increase in Chlorella strains calorific values when grown in low nitrogen medium. *Enzyme & Microbiol Technology* 27, 631–635.

Institute of Electrical Engineers (IEE) (2002) *The Environmental Effects of Electricity Generation*. Institute of Electrical Engineers, London.

Intergovernmental Panel on Climate Change (IPCC) (1996) *Climate Change 1995: Impacts, Adaptation and Mitigation of Climate Change*. Available at http://www.ipcc.ch

Intergovernmental Panel on Climate Change (IPCC) (2006) *Carbon Dioxide Capture and Storage*. Available at http://www.ipcc.ch

Intergovernmental Panel on Climate Change (IPCC) (2007) *Climate Change 2007: The Physical Science Basis*. Available at http://www.ipcc.ch

International Energy Agency (IEA) (2002) *Renewables in Global Energy Supply*. Paris, France. Available at http://www.iea.org

International Energy Agency (IEA) (2005a) *World Energy Outlook 2005*. Available at http://www.worldenergyoutlook.org

International Energy Agency (IEA) (2005b) *Biofuels for Transport*. Available at http://www.iea.org

International Energy Agency (IEA) (2007) *Energy Statistics*. Available at http://www.iea.org

International Energy Agency (IEA) (2008a) *IEA Bioenergy Update 29*. Available at http://www.iea.org

International Energy Agency (IEA) (2008b) *Energy Statistics*. Available at http://www.iea.org

Jager-Waldau, A. (2007) Photovoltaics and renewable energies in Europe. *Renewable & Sustainable Energy Reviews* 11, 1414–1437.

Johansson, D.J.A. and Azar, C. (2007) A scenario based analysis of land competition between food and bioenergy production in the US. *Climate Change* 82, 267–291.

Joint Research Centre, European Commission (2007) Well-to-wheels analysis of future automotive fuels and powertrains in the European context. Available at http://ies.jrc.ec.europa.eu/WTW

Kalam, M.A. and Masjuki, H.H. (2002) Biodiesel from palmoil – an analysis of its properties and potential. *Biomass Bioenergy* 23, 471–479.

Kalligeros, S., Zannikos, F., Stournas, S., Lois, E., Anastopoulos, G., Teas, C. and Sakellaropoulos, F. (2003) An investigation of using biodiesel/marine diesel blends on the performance of a stationary diesel engine. *Biomass Bioenergy* 24, 141–149.

Kapdan, I.K. and Kargi, F. (2006) Bio-hydrogen production from waste materials. *Enzyme & Microbial Technology* 38, 569–582

Karinen, R.S. and Krause, A.O.I. (2006) New biocomponents from glycerol. *Applied Catalysis A* 306, 128–133.

Karmee, S.K. and Chadha, A. (2005) Preparation of biodiesel from crude oil of *Pongamia pinnata*. *Bioresource Technology* 96, 1425–1429.

Kebede-Westhead, E., Pizarro, C. and Mulbry, W.W. (2006) Treatment of swine manure effluent using freshwater algae: production, nutrient recovery, and elemental composition of algal biomass at four effluent loading rates. *Journal of Applied Phycology* 18, 41–46.

Kegl, B. (2008) Effects of biodiesel on emissions of a bus diesel engine. *Bioresource Technology* 99, 863–873.

Kenisarin, M. and Mahkamov, K. (2007) Solar energy storage using phase change materials. *Renewable & Sustainable Energy Reviews* 11, 1913–1965.

Keoleian, G. and Volk, T. (2005) Renewable energy from willow biomass crops: life cycle energy, environmental and economic performance. *Critical Review of Plant Science* 24, 385–406.

Kim, S. and Dale, B.E. (2005) Life cycle assessment of various cropping systems utilized for producing biofuels: bioethanol and biodiesel. *Biomass Bioenergy* 29, 426–439.

Kirschbaum, M.U.F. (2003) To sink or burn? A discussion of the practical contributions of forest to greenhouse gas balances through storing carbon or providing biofuels. *Biomass Bioenergy* 24, 297–310.

Knauer, J. and Southgate, P.C. (1999) A review of the nutritional requirements of bivalves and the development of alternative and artificial diets for bivalve aquaculture. *Review of Fisheries Science* 7, 241–280.

Knothe, G. (2001) Historical perspectives on vegetable oil-based diesel fuels. *Inform* 12, 1103–1107.

Knothe, G. (2005) Dependence of biodiesel fuel properties on the structure of fatty acid alkyl esters. *Fuel Processing Technology* 86, 1059–1070.

Knothe, G., Matheaus, A.C. and Ryan, III T.W. (2003) Cetane numbers of branched and straight-chain fatty esters determined in an ignition quality tester. *Fuel* 82, 971–975.

Kvenvolden, K.A. (1999) Potential effects of gas hydrates on human welfare. *Proceedings of the National Academy of Sciences (USA)* 96, 3420–3426.

Labeckas, G. and Slavinskas, S. (2006) Performance of direct-injection off-road diesel engine on rapeseed oil. *Renewable Energy* 31, 849–863.

Laherrere, J. (2001) Forecasting future production from past discovery. OPEC Seminar. Available at http://www.opec.org

Lang, X., Dalai, A.K., Bakhshi, N.N., Reaney, M.J. and Hertz, P.B. (2001) Preparation and characterization of bio-diesels from various bio-oils. *Bioresource Technology* 80, 53–62.

Lee, S.Y. and Holder, G.D. (2001) Methane hydrates potential as a future energy source. *Fuel Processing Technology* 71, 181–186.

Lettens, S., Muys, B., Ceulemans, R., Moons, E., Garcia, J. and Coppin, P. (2003) Energy budget and greenhouse gas balance evaluation of sustainable coppice systems for electricity production. *Biomass Bioenergy* 24, 179–197.

Leung, D.Y.C. and Guo, Y. (2006) Transesterification of neat and used frying oil: optimization for biodiesel production. *Fuel Processing Technology* 87, 883–890.

Lewandowski, I., Sculock, J.M.O., Lindvall, E. and Christou, M. (2003) The development and current status of perennial rhizomatous grasses as energy crops in the US and Europe. *Biomass Bioenergy* 25, 335–361.

Li, H.B., Chen, F., Zhang, T.Y., Yang, F.Q. and Xu, G.Q. (2001) Preparative isolation and purification of lutein from the microalga *Chlorella vulgaris* by high-speed counter-current chromatography. *Journal of Chromatography A* 905, 151–155.

Liang, Y.C., May, C.Y., Foon, C.S., Ngan, M.A., Hock, C.C. and Basiron, Y. (2006) The effect of natural and synthetic antioxidants on the oxidative stability of palm diesel. *Fuel* 85, 867–870.

Licht, F.O. (2006) *FO Licht's World Ethanol & Biofuels Report*, Vol. 4, January 2006, F.O. Licht.

Linko, Y.Y., Lamsa, M., Wu, X., Uosukainen, W., Sappala, J. and Linko, P. (1998) Biodegradable products by lipase biocatalysts. *Journal of Biotechnology* 66, 41–50.

Lorenz, D. and Morris, D. (1995) How Much Energy Does It Take to Make a Gallon of Ethanol? Institute for Local Self-Reliance, Washington, DC.

Louwrier, A. (1998) Biodiesel: tomorrow's gold. *Biologist* 45, 1721–1728.

Ma, F., Clements, L.D. and Hanna, M.A. (1998) The effects of catalyst, free fatty acids, and water on transesterification of beef tallow. *Transactions of the ASAE* 41, 1261–1264.

Ma, F. and Hanna, M.A. (1999) Biodiesel production: a review. *Bioresource Technology* 70, 1–15.

Machacon, H.T.C., Shiga, S., Karasawa, T. and Nakamura, H. (2001) Performance and emission characteristics of a diesel engine fueled with coconut oil–diesel fuel blend. *Biomass Bioenergy* 20, 63–69.

Maeda, K., Owada, M., Kimura, N., Omata, K. and Karube, I. (1995) CO_2 fixation from the flue gas on coal-fired thermal power plant by microalgae. *Energy Conversion & Management* 35, 717–720.

Makareviciane, V. and Janulis, P. (2003) Environemntal effect of rapeseed oil ethyl ester. *Renewable Energy* 28, 2395–2403.

Malca, J. and Freire, F. (2006) Renewability and life-cycle energy efficiency of bioethanol and bio-ethyl tertiary butyl ether (bioETBE): assessing the implications of allocation. *Energy* 31, 3362–3380.

Mathews, J. (2007) Seven steps to curb global warming. *Energy Policy* 35, 4247–4259.

Matsumoto, H., Shioji, N., Hanasaki, A., Ikuta, Y., Fukuda, Y., Sato, M., Endo, N. and Tsukamoto, T. (1995) Carbon dioxide fixation by microalgae photosynthesis using actual flue gas discharged from a boiler. *Applied Biochemistry & Biotechnology* 51/52, 681–692.

Matthews, R.W. (2001) Modeling of energy and carbon budgets of wood fuel coppice systems. *Biomass Bioenergy* 21, 1–19.

McEldowney, J.F., Hardman, D.J. and Waite, S. (1993) *Pollution, Ecology and Biotreatment.* Longman Scientific & Technical, Harlow, UK.

McLaren, J.S. (2005) Crop biotechnology provides an opportunity to develop a sustainable future. *Trends in Biotechnology* 23, 339–342.

Meher, L.C., Sagar, D.V. and Naik, S.N. (2006a) Technical aspects of biodiesel production by transesterification – a review. *Renewable & Sustainable Energy Reviews* 10, 248–268.

Meher, L.C., Kulkarni, M.G., Dalai, A.K. and Naik, S.N. (2006b) Transesterification of karanja (*Pongamia pinnata*) oil by solid basic catalysis. *European Journal of Lipid Science and Technology* 108, 389–397.

Meher, L.C., Dharmagadda, S.S. and Naik, S.N. (2006c) Optimization of alkali-catalysed transesterification of *Pongamia pinnata* oil for production of biodiesel. *Bioresource Technology* 97, 1392–1397.

Melaina, M.W. (2007) Turn of the century refueling: a review of innovations in early gasoline refueling methods and analogies for hydrogen. *Energy Policy* 35, 4919–4934.

Mendes, R.L., Nobre, B.P., Cardoso, M.T., Pereira, A.P. and Palanra, A.F. (2003) Supercritical carbon dioxide extraction of compounds with pharmaceutical importance from microalgae. *Inorganica Chimica Acta* 356, 328–334.

Miao, X. and Wu, Q. (2006) Biodiesel production from heterotrophic micralgal oil. *Bioresource Technology* 97, 841–846.

Midilli, A., Ay, M., Dincer, I. and Rosen, M.A. (2005) On hydrogen and hydrogen energy strategies I: current status and needs. *Renewable & Sustainable Energy Reviews* 9, 255–271.

Minowa, T., Yokoyama, S.Y., Kishimoto, M. and Okakura, T. (1995) Oil production from algal cells of *Dunaliella tertiolecta* by direct thermochemical liquefaction. *Fuel* 74, 1735–1738.

Miron, A.S., Gomez, A.C., Camacho, F.G., Molina Grima, E. and Chisti, Y. (1999) Comparative evaluation of compact photobioreactors for large-scale monoculture of microalgae. *Journal of Biotechnology* 70, 249–270.

Mittelbach, M. (1990) Lipase catalysed alcoholysis of sunflower oil. *Journal of American Oil Chemists' Society* 67, 168–170.

Mittelbach, M. and Gangl, S. (2001) Long storage stability of biodiesel made from rapeseed and used frying oil. *Journal of American Oil Chemists' Society* 78, 573–577.

Molina Grima, E., Perez, J.A.S., Camacho, F.G. and Alonso, D.L. (1993) n-3 PUFA productivity in chemostat cultures of microalgae. *Applied Microbiology Biotechnology* 38, 599–605.

Molina Grima, E., Robles, M.A., Gimenez Gimenez, A., Perez, J.A., Garcia Camacho, F.G. and Sanchez, J.L. (1994) Comparison between extraction of lipids and fatty acids from microalgal biomass. *Journal of American Oil Chemists' Society* 71, 955–959.

Molina Grima, E., Fernandez, J., Acien Fernanadez, F.G. and Chisti, Y. (2001) Tubular photobioreactor design for algal cultures. *Journal of Biotechnology* 92, 113–131.

Molina Grima, E., Belarbi, E.H., Acien Fernandez, F.G., Medina, A.R. and Chisti, Y. (2003) Recovery of microalgal biomass and metabolites: process options and economics. *Biotechnology Advances* 20, 491–515.

Monyem, A. and van Gerpen, J.H. (2001) The effect of biodiesel oxidation on engine performance and emissions. *Biomass Bioenergy* 20, 317–325.

Morohoshi, N. and Kajita, S. (2001) Formation of a tree having a low lignin content. *Journal of Plant Research* 114, 517–523.

Mortimer, N.D., Cormack, P., Elsayed, M.A. and Horne, R.E. (2003) Evaluation of the Comparative Energy, Global Warming and Socio-Economic Costs and Benefits of Biodiesel. DEFRA Report. Available at http://www.defra

Mu, Y., Teng, H., Zhang, D.-J., Wang, W. and Xiu, Z.-L. (2006) Microbial production of 1,3-propanediol by *Klebsiella pneumoniae* using crude glycerol from biodiesel preparations. *Biotechnology Letters* 28, 1755–1759.

Murillo, S., Miguez, J.L., Porteiro, J., Granada, E. and Moran, J.C. (2007) Performance and exhaust emissions in the use of biodiesel in outboard diesel engines. *Fuel* 86, 1765–1771.

Nagle, N. and Lemke, P. (1990) Production of methyl-ester fuel from microalgae. *Applied Microbiology and Biotechnology* 24, 355–361.

National Renewable Energy Laboratory (NREL) (1998) Nonpetroleum based fuels. Available at http://www.nrel.gov

National Society for Clean Air and Environmental Protection (NSCA) (2006) Biogas as a road transport fuel. Available at http://www.nsca.org.uk

Nebel, B.A. and Mittelbach, M. (2006) Biodiesel from extracted fat out of meat and bone meal. *European Journal of Lipid Science and Technology* 108, 398–403.

Ni, M., Leung, D.Y.C., Leung, M.K.H. and Sumathy, K. (2006) An overview of hydrogen production from biomass. *Fuel Processing Technology* 87, 461–472.

Niven, R.K. (2005) Ethanol in gasoline: environmental impacts and sustainability review article. *Renewable & Sustainable Energy Reviews* 9, 535–555.

Nwafor, O.M.I. (2004) Emission characteristics of diesel engine operating on rapeseed methyl ester. *Renewable Energy* 29, 119–129.

Nwafor, O.M.I. and Rice, G. (1995) Performance of rapeseed methyl ester in diesel engine. *Renewable Energy* 6, 335–342.

Odell, P.R. (1999) Dynamics of energy technologies and global change. *Energy Policy* 27, 737–742.

OECD (2007) *Sustainable Development*. OECD Publications, Paris.

Ogbonna, J.C., Yada, H., Masui, H. and Tanaka, H. (1996) A novel internally illuminated stirred tank photobioreactor for large-scale cultivation of photosynthetic cells. *Journal of Fermentation & Bioengineering* 82, 61–67.

Ogden, J.M., Steinbugler, M.M. and Kreutz, T.G. (1999) A comparison of hydrogen, methanol and gasoline as fuels for fuel cell vehicles: implications for vehicle design and infrastructure development. *Journal of Power Sources* 79, 143–168.

Olquin, E.J., Galicia, S., Camacho, R., Mercado, G. and Perez, T.J. (1997) Production of *Spirulina* sp in sea water supplemented with anaerobic effluents in outdoor raceways under temperate climatic conditions. *Applied Microbiology and Biotechnology* 48, 242–247.

Ono, E. and Cuello, J.L. (2006) Feasibility assessment of microalgal carbon dioxide sequestration technology with photobioreactor and solar collector. *Biosystems Engineering* 95, 597–606.

Ooi, Y.S., Zakaria, R., Mohamed, A.R. and Bhatia, S. (2004) Catalytic conversion of palm oil-based fatty acid mixture to liquid fuel. *Biomass Bioenergy* 27, 477–484.

Openshaw, K. (2000) A review of *Jatropha curcas*: an oil plant of unfulfilled promise. *Biomass Bioenergy* 19, 1–15.

Papanikolaou, S., Fakas, S., Fick, M., Chevalot, I., Galiotou-Panayotou, M., Komaitis, M., Marc, I. and Aggelis, G. (2008) Biotechnological valorisation of raw glycerol discharged after biodiesel (fatty acid methyl esters) manufacturing process: production of 1,3-propanediol, citric acid and single cell oil. *Biomass Bioenergy* 32, 60–71.

Parikka, M. (2004) Global biomass fuel resources. *Biomass Bioenergy* 27, 613–620.

Pereira, M.G. and Mudge, S.M. (2004) Cleaning oiled shores: laboratory experiments testing the potential use of vegetable oil biodiesels. *Chemosphere* 54, 297–304.

Peterson, G.R. and Scarrach, W.P. (1984) Rapeseed oil transesterification by heterogeneous catalysis. *Journal of American Oil Chemists' Society* 61, 1593–1596.

Peterson, C.L. and Hustrulid, T. (1998) Carbon cycle for rapeseed oil biodiesel fuels. *Biomass Bioenergy* 14, 91–101.

Peterson, C.L., Reece, D.L., Thompson, J.C., Beck, S.M., Chase, C. (1996) Ethyl ester of rapeseed used as a biodiesel fuel: a case study. *Biomass Bioenergy* 10, 331–336.

Pimental, D. (1991) Ethanol fuels: energy security, economics, and the environment. *Journal of Agricultural & Environmental Ethics* 4, 1–13.

Pimental, D. (2004) The limits of biomass energy. *Encyclopedia of Physical Sciences and Technology* 159–171.

Poole, M. and Towler, G. (1989) Alcohol in Brazil. *The Chemical Engineer* May 48–50.

Poullikkas, A. (2005) An overview of current and future sustainable gas turbine technologies. *Renewable & Sustainable Energy Reviews* 9, 409–443.

Powlson, D.S., Riche, A.B. and Shield, I. (2005) Biofuels and other approaches for decreasing fossil fuel emissions from agriculture. *Annals of Applied Biology* 146, 193–201.

Pradeep, V. and Sharma, R.P. (2007) Use of HOT EGR for NOx control in a compression ignition engine fuelled with bio-diesel from Jatropha oil. *Renewable Energy* 32, 1136–1154.

Prasad, S., Singh, A. and Joshi, H.C. (2007) Ethanol as an alternative fuel from agricultural, industrial and urban residues. *Resources Conservation & Recycling* 50, 1–39.

Prins, M.J., Ptasinski, K.J. and Janssen, F.J.J.G. (2004) Exergetic optimization of a production process of Fischer-Tropsch fuels from biomass. *Fuel Processing Technology* 86, 375–389.

Puhan, S., Vedaraman, N., Ram, B.V.B., Sankarnarayanan, G. and Jeychandran, K. (2005) Mahua oil (*Madhuca indica* seed oil) methyl ester as biodiesel: preparation and emission characteristics. *Biomass Bioenergy* 28, 87–93.

Quadrelli, R. and Peterson, S. (2007) The energy-climate challenge: recent trends in CO_2 emissions from fuel combustion. *Energy Policy* 35, 5938–5952.

Raheman, H. and Phadatare, A.G. (2004) Diesel engine emissions and performance from blends of karanja methyl ester and diesel. *Biomass Bioenergy* 27, 393–397.

Rajagopalan, S., Datar, R.P. and Lewis, R.S. (2002) Formation of ethanol from carbon monoxide via a new microbial catalyst. *Biomass Bioenergy* 23, 487–493.

Rakopoulos, C.D., Antonopoulos, K.A., Rakopoulos, D.T., Hountalas, D.T. and Giakoumis, E.G. (2006) Comparative performance and emissions study of a direct injection diesel engine using blends of diesel fuel with vegetable oils or bio-diesels of various origins. *Energy Conversion & Management* 47, 3272–3287.

Ramadhas, A.S., Jayaraj, S. and Muraleedharan, C. (2005) Biodiesel production from high FFA rubber seed oil. *Fuel* 84, 335–340.

Reijnders, L. (2006) Conditions for sustainablility of biomass based fuel use. *Energy Policy* 34, 863–876.

Reijnders, L. (2008) Do biofuels from microalgae beat biofuels from terrestrial plants? *Trends in Biotechnology* 26, 349–350.

Reyes, J.F. and Sepulveda, M.A. (2006) PM-10 emissions and power of a diesel engine fueled with crude and refined biodiesel from salmon oil. *Fuel* 85, 1714–1719.

Richmond, A., Boussiba, S., Vaonshak, A. and Kopel, R. (1993) A new tubular reactor for mass production of microalgae outdoors. *Journal of Applied Phycology* 5, 327–332.

Robert, M., Hulten, P. and Frostell, B. (2007) Biofuels in the energy transition beyond peak oil: a macroscopic study of energy demand in the Stockholm transport system 2030. *Energy* 32, 2089–2098.

Roberts, M.C. (2007) E85 and fuel efficiency: an empirical analysis of 2007 EPA test data. *Energy Policy* 36, 1233–1235.

Roberts, L.E.J., Liss, P.S. and Saunders, P.A.H. (1990) *Power Generation and the Environment.* Oxford University Press, Oxford.

Rosenberg, J.N., Oyler, G.A., Wilkinson, L. and Betenbaugh, M.J. (2008) A green light for engineered algae: redirecting metabolism to fuel a biotechnology revolution. *Current Opinion in Biotechnology* 19, 430–436.

Rosenberger, A., Kaul, H.P., Senn, T. and Aufhammer, W. (2002) Costs of bioethanol production from winter cereals: the effect of growing conditions and crop production intensity levels. *Industrial Crops & Products* 15, 91–102.

Rowe, R.L., Street, N.R. and Taylor, G. (2009) Identifying potential environmental impacts of large-scale deployment of dedicated bioenergy crops in the UK. *Renewable & Sustainable Energy Reviews* 13, 271–290.

Royon, D., Daz, M., Ellenrieder, G. and Locatelli, S. (2007) Enzymatic production of biodiesel from cotton seed oil using n-butanol as a solvent. *Bioresource Technology* 98, 648–653.

Rukes, B. and Taud, R. (2004) Status and perspectives of fossil power generation. *Energy* 29, 1853–1874.

Ryan, L., Convery, F. and Ferreira, S. (2006) Stimulating the use of biofuels in the European Union: implications for climate change policy. *Energy Policy* 34, 3184–3194.

Sahoo, P.K., Das, L.M., Babu, M.K.G. and Naik, S.N. (2006) Biodiesel development from high acid value polanga seed oil and performance evaluation in a CI engine. *Fuel* 86, 448–454.

Saka, S. and Kusdrana, D. (2001) Biodiesel fuel from rapeseed oil as prepared in supercritical methanol. *Fuel* 80, 225–231.

Sami, M., Annamalai, K. and Wooldridge, M. (2001) Co-firing of coal and biomass fuel blends. *Progress in Energy & Combustion Science* 27, 171–214.

Sarin, R., Sharma, M., Sinharay, S. and Malhotra, R.K. (2007) Jatropha-palm biodiesel blends: an optimum mix for Asia. *Fuel* 86, 1365–1371.

Sato, T., Usui, S., Tsuchiya, Y. and Kondo, Y. (2006) Invention of outdoor closed type photobioreactor for microalgae. *Energy Conversion & Management* 47, 791–799.

Sawayama, S., Dote, S.I.Y. and Yokoyama, S.Y. (1995) CO_2-fixation and oil production through microalga. *Energy Conversion & Management* 36, 729–731.

Schober, S., Seidl, I. and Mittelbach, M. (2006) Ester content evaluation in biodiesel from animal fats and lauric oils. *European Journal of Lipid Science and Technology* 108, 309–314.

Scragg, A.H. (2005) *Environmental Biotechnology*. Oxford University Press, Oxford.

Scragg, A.H., Illman, A.M., Carden, A. and Shales, S.W. (2002) Growth of microalgae with increased calorific values in a tubular bioreactor. *Biomass Bioenergy* 23, 67–73.

Scragg, A.H., Morrison, J. and Shales, S.W. (2003) The use of a fuel containing *Chlorella vulgaris* in a diesel engine. *Enzyme & Microbial Technology* 33, 884–889.

Selmi, B. and Thomas, D. (1998) Immobilized lipase-catalyzed ethanolysis of sunflower oil in solvent-free medium. *Journal of American Oil Chemists' Society* 75, 691–695.

Semelsberger, T.A., Borup, R.L. and Greene, H.L. (2006) Dimethyl ether (DME) as an alternative fuel. *Journal of Power Sources* 156, 497–511.

Sendzikiene, E., Makarevicene, V., Janulis, P. and Makarevicute, D. (2007) Biodegradability of biodiesel fuel of animal and vegetable origin. *European Journal of Lipid Science and Technology* 109, 493–497.

Shakya, B.D., Aye, L. and Musgrave, P. (2005) Technical feasibility and financial analysis of hybrid wind-photovoltaic system with hydrogen storage for Cooma. *International Journal of Hydrogen Energy* 30, 9–20.

Shapouri, H., Duffield, J.A. and Wang, M. (2002) The Energy Balance of Corn Ethanol: An Update. US Department of Agriculture, Report 813.

Shay, E.G. (1993) Diesel fuel from vegetable oils: status and opportunities. *Biomass Bioenergy* 4, 227–242.

Sheehan, J., Dunahay, T., Benemann, J. and Roessler, P. (1998) A look back at the US Department of Energy's aquatic species program: biodiesel from algae. NREL/TP-580–24190.

Shiuya, I., Tamura, G., Ishikawa, T. and Hara, S. (1992) Cloning of the α-amylase cDNA of *Aspergillus shirousamii* and its expression in *Saccharomyces cerevisiae*. *Bioscience Biotechnology Biochemistry* 56, 174–179.

Siler-Marinkovic, S. and Tomasevic, A. (1998) Transesterification of sunflower oil *in situ*. *Fuel* 77, 1389–1391.

Skjanes, K., Lindblad, P. and Muller, J. (2007) BioCO_2-multidisciplinary, biological approach using solar energy to capture CO_2 while producing H_2 and high value products. *Biomolecular Engineering* 24, 405–413.

Smeets, E.M.W. and Faaij, A.P.C. (2007) Bioenergy potentials from forestry in 2050. *Climate Change* 81, 353–390.

Solomon, B.D., Barnes, J.R. and Halvorsen, K.E. (2007) Grain and cellulosic ethanol: history, economics, and energy policy. *Biomass Bioenergy* 31, 416–425.

Soriano, N.U., Migo, V.P. and Matsumura, M. (2005) Ozonized vegetable oil as pour point depressant for neat biodiesel. *Fuel* 85, 25–31.

Spolaore, P., Joannis-Cassan, C., Duran, E. and Isambert, A. (2006) Commercial applications of microalgae. *Journal of Bioscience and Bioengineering* 101, 87–96.

Sreeprasanth, P.S., Srivastava, R., Srinivas, D. and Ratnasamy, P. (2006) Hydrophobic, solid acid catalysts for production of biofuels and lubricants. *Applied Catalysis A General* 314, 148–159.

Srivastava, A. and Prasad, R. (2000) Triglyceride-based diesel fuels. *Renewable & Sustainable Energy Reviews* 4, 111–133.

Stambouli, A.B. and Traversa, E. (2002) Fuels cells, an alternative to standard sources of energy. *Renewable & Sustainable Energy Reviews* 6, 297–306.

Stanhope-Seta (2007) Biofuel specifications. Available at http://www.biofueltesting.com

Stern, N. (2006) *Stern review on the economics of climate change*. HM Treasury.

Strahan, D. (2007) *Last Oil Shock: Oil Depletion Atlas*. Available at http://www.lastoilshock.com

Suppes, G.J., Mohanprasad, A.D., Doskocil, E.J., Mankidy, P.J. and Goff, M.J. (2004) Trans-esterification of soybean oil with zeolite and metal catalysts. *Applied Catalysis A* 257, 213–223.

Sun, Y. and Cheng, J. (2002) Hydrolysis of lignocellulosic materials for ethanol production: a review. *Bioresource Technology* 83, 1–11.

Thelen, J.J. and Ohlrogge, J.B. (2002) Metabolic engineering of fatty acid biosynthesis in plants. *Metabolic Engineering* 4, 12–21.

Themelis, N.J. and Ulloa, P.A. (2007) Methane generation in landfills. *Renewable Energy* 32, 1243–1257.

Tomasevic, A.V. and Siler-Marinlovic, S.S. (2003) Methanolysis of used frying oil. *Fuel Processing Technology* 81, 1–6.

Travieso, L., Pellon, A., Benitez, F., Sanchez, E., Borja, R., O'Farrill, N. and Weiland, P. (2002) BIOALGA reactor: preliminary studies for heavy metals removal. *Biochemical Engineering* 12, 87–91.

Tredici, M.R. and Zittelli, G.C. (1998) Efficiency of sunlight utilization: tubular versus flat photo-bioreactors. *Biotechnology and Bioengineering* 57, 187–197.

UK Report to European Commission Article 4 of the Biofuels Directive (2003/30/EC) UK Department for Transport, 2006.

Ulusoy, Y., Tekin, Y., Cetinkaya, M. and Karaosmanoglu, F. (2004) The engine tests of bio-diesel from used frying oil. *Energy Sources* 26, 927–932.

United Nations Environment Programme (UNEP) (2000) World energy assessment. United Nations Development Programme, New York. Available at http://www.unep.ch

United Nations Environment Programme (UNEP) (2006) Vital climate changes graphics. Available at http://www.unep.ch

United Nations Framework Convention on Climate Change (UNFCCC) (2002) A guide to the climate change convention process. Available at http://www.unfccc.int

United Nations Framework Convention on Climate Change (UNFCCC) (2008) Counting emissions and removals. Available at http://www.unfccc.int

Upreti, B.R. and van der Horst, D. (2004) National renewable energy policy and local opposition in the UK: the failed development of a biomass electricity plant. *Biomass Bioenergy* 26, 61–69.

Usta, N. (2005) Use of tobacco seed oil ester in a turbocharged indirect injection diesel engine. *Biomass Bioenergy*. 28, 77–86.

Van der Drift, A. and Boerringter, H. (2006) Synthesis gas from biomass. SYNBIOS. ECN-C-06–001.

Van der Linden, S. (2006) Bulk energy storage potential in the USA, current developments and future prospects. *Energy* 31, 3446–3457.

Van Gerpen, J. (2005) Biodiesel processing and production. *Fuel Processing Technology* 86, 1097–1107.

Van Maris, A.J.A., Abbott, D.A., Bellissimi, E., van den Brink, J., Kuyper, M., Luttik, M.A.H., Wisselink, H.W., Scheffers, W.A., van Dijken, J.P. and Pronk, J.T. (2006) Alcoholic fermentation of carbon sources in biomass hydrolysates by *Saccharomyces cerevisiae*: current status. *Antonie van Leeuwenhoek* 90, 391–418.

Van Mierlo, J., Maggetto, G. and Lataire, P.H. (2006) Which energy source for road transport in the future? A comparison of battery, hybrid and fuel cell vehicles. *Energy Conservation & Management* 47, 2748–2760.

Venturi, P. and Venturi, G. (2003) Analysis of energy comparison for crops in European agricultural systems. *Biomass Bioenergy* 25, 235–255.

Verge, X.P.C., Kimpe, C. De and Desjardins RL (2007) Agricultural production, greenhouse gas emissions and mitigation potential. *Agricultural Forest Meterology* 142, 255–269.

Vincente, G., Martinez, M. and Aracil, J. (2004) Integrated biodiesel production: a comparison of different homogeneous catalyst systems. *Bioresource Technology* 92, 297–305.

Walker, R.L., Walker, K.C. and Booth, E.J. (2003) Adaptation potential of the novel oilseed crop, Honesty (*Lunaria annua* L), to the Scottish climate. *Industrial Crops & Products* 18, 7–15.

Wang, M., Saricks, C. and Santini, D. (1999) Effects of Fuel Ethanol Use on Fuel-Cycle Energy and Greenhouse Gas Emissions. US Department of Energy, Argonne National Laboratory, Center for Transportation Research, Argonne IL.

Wang, Y., Ou, S., Liu, P. and Zhang, Z. (2006) Preparation of biodiesel from waste cooking oil via two-step catalyzed process. *Energy Conversion & Management* 48, 184–181.

Wardle, D.A. (2003) Global sale of green air travel supported using biodiesel. *Renewable & Sustainable Energy Reviews* 7, 1–64.

Watanabe, Y. and Hall, D.O. (1996) Photosynthetic production of the filamentous cyanobacterium *Spirulina platensis* in a cone-shaped helical tubular photobioreactor. *Applied Microbiology and Biotechnology* 44, 693–698.

Watanabe, Y., Shimada, Y., Sugihara, A. and Tominaga, T. (2002) Conversion of degummed soybean oil to biodiesel fuel with immobilized *Candida antarctica* lipase. *Journal of Molecular Catalysis B: Enzyme* 17, 151–155.

Watkins, R.S., Lee, A.F. and Wilson, K. (2004) Li-CaO catalysed tri-glyceride transesterification for biodiesel applications. *Green Chemistry* 6, 335–340.

Wheals, A.E., Basso, L.C., Alves, D.M.G. and Amorim, H.V. (1999) Fuel ethanol after 25 years. *Trends in Biotechnology* 17, 482–487.

Williamson, A.M. and Badr, O. (1998). Assessing the viability of using rape methyl ester (RME) as an alternative to mineral diesel fuel for powering road vehicles in the UK. *Applied Energy* 59, 187–214.

Winter, C.J. (2005) Into the hydrogen energy economy: milestones. *International Journal of Hydrogen Energy* 30, 681–685.

Woodfuel (2007) Available at http://www.woodfuel.org.uk

World Resources Institute (2006) Navigating the numbers: a journalist's guide. Available at http://www.wri.org

Wright, L. (2006) Worldwide commercial development of bioenergy with a focus on energy crop-based projects. *Biomass Bioenergy* 30, 706–714.

Wu, W.H., Fogia, A., Marmer, W.N. and Phillips, J.G. (1999) Optimizing production of ethyl esters of grease using 95% ethanol by response surface methodology. *Journal of American Oil Chemists' Society* 76, 517–521.

Wuebbles, D.J., Jain, A., Edmonds, J., Harvey, D. and Hayhoe, K. (1999). Global change: state of the science. *Environmental Pollution* 26, 157–168.

Xu, H., Miao, X. and Wu, Q. (2006) High quality biodiesel production from a microalga *Chlorella protothecoides* by heterotrophic growth in fermenters. *Journal of Biotechnology* 126, 499–507.

Yazdani, S.S. and Gonzalez, R. (2007) Anaerobic fermentation of glycerol: a path to economic viability for biofuels industry. *Current Opinion in Biotechnology* 18, 1–7.

Yoshihara, K., Nagase, H., Eguchi, K., Hirata, K. and Miyamoto, K. (1996) Biological elimination of nitric oxide and carbon dioxide from flue gas by marine microalga NOA-113 cultivated in a long tubular photobioreactor. *Journal of Fermentation and Bioengineering* 82, 351–354.

Zeiler, K.G., Heacox, D.A., Toon, S.T., Kadam, K.L. and Brown, L.M. (1995) The use of microalgae for assimilation and utilization of carbon dioxide from fossil fuel-fired power plant fuel gas. *Energy Conversion & Management* 36, 707–712.

Zhang, K., Kurano, N. and Miyachi, S. (2002) Optimized aeration by carbon dioxide gas for microalgal production and mass transfer characterization in a vertical flat-plate photobioreactor. *Bioprocess and Biosystems Engineering* 25, 97–101.

Zhang, O., Chang, J., Wang, T. and Xu, Y. (2007) Review of biomass pyrolysis oil properties and upgrading research. *Energy Conservation & Management* 48, 87–92.

Zhang, X., Peterson, C.L., Reece, D., Moller, G. and Haws, R. (1998) Biodegradability of bio-diesel in the aquatic environment. *Transactions of the ASAE* 41, 1423–1430.

Zhang, Y., Dube, M.A., McLean, D.D. and Kates, M. (2003) Biodiesel production from waste cooking oil: 1 – process design and technological assessment. *Bioresource Technology* 89, 1–16.

Zhou, L. (2005) Progress and problems in hydrogen storage methods. *Renewable & Sustainable Energy Reviews* 9, 395–408.

Zhu, H., Wu, Z., Chen, Y., Zang, P., Duan, S., Liu, X. and Mao, Z. (2006) Preparation of bio-diesel catalysed by solid super base of calcium oxide and its refining process. *Chinese Journal of Catalysis* 27, 391–396.

Zittelli, G.C., Rodolfi, L. and Tredici, M.R. (2003) Mass cultivation of *Nannochloropsis* sp in annular reactors. *Journal of Applied Phycology* 15, 107–114.

Zubr, J. (1997) Oil-seed crop: *Camelina sativa. Industrial Crops & Products* 6, 113–119.

Zullaikah, S., lai, C.C., Vali, S.R. and Ju, Y.H. (2005) A two-step acid catalysed process for the production of biodiesel from rice bran oil. *Bioresource Technology* 96, 1889–1896.

Index

Page numbers in **bold** type refer to figures and tables.

Vacuum ethanol extraction 125–126, **126**
Vehicles
    battery electric 48–49
    engine performance, with
        biofuels 169–173, **172**
    hybrid 49
    hydrogen-powered 98, 100–101, **101**
    modification for alternative fuels 89,
        101, 102, 109, **109**
    use of fuel blends 106, 108–109
Venezuela, heavy oil reserves 12

Waste, animal and municipal 55, 72, 89, 93
Waste cooking oil, source of biodiesel 153,
    160, 163, **164**
Water power, pre-industrial use 1–2
Water-shift reaction 51, 93, 94, 97
Wave power 7, 57
Well to-wheel (WTW) analysis 183–184,
    185, 206–210, **211**

Willow, use as biofuel 66
Wind power 57, 60
    pre-industrial use 1–2
    siting of farms 7
Wood biomass 1, 65, 76–77, 195–196
World Climate Conference (1979,
    Geneva) 21

Xylose catabolism 200, 201, **202**

Yeasts
    alternatives to *Saccharomyces*
        cerevisiae **117**, 125, 201
    sugar fermentation 114–117, **116,**
        124–125, **202**

Zeldovitch mechanism
    (N/O reaction) 30